Federated AI for Real-World Business Scenarios

Dinesh C. Verma

IBM Fellow and Department Group Manager
Distributed AI, IBM TJ Watson Research Center
Yorktown Heights, NY, USA

CRC Press
Taylor & Francis Group
Boca Raton London New York

CRC Press is an imprint of the
Taylor & Francis Group, an **informa** business

A SCIENCE PUBLISHERS BOOK

First edition published 2021
by CRC Press
6000 Broken Sound Parkway NW, Suite 300, Boca Raton, FL 33487-2742

and by CRC Press
2 Park Square, Milton Park, Abingdon, Oxon, OX14 4RN

© 2021 Taylor & Francis Group, LLC

CRC Press is an imprint of Taylor & Francis Group, LLC

Library of Congress Cataloging-in-Publication Data

Names: Verma, Dinesh C., author.
Title: Federated AI for real-world business scenarios / Dinesh Verma, IBM
    Fellow and Department Group Manager, Distributed AI, IBM TJ Watson
    Research Center Yorktown Heights, NY, USA.
Description: First edition. | Boca Raton, FL : CRC Press, 2021. | Includes
    bibliographical references and index. | Summary: "This book provides a
    holistic overview of all aspects of federated learning, which allows
    creation of real-world applications in contexts where data is dispersed
    in many different locations. It covers all stages in the creation and
    use of AI based applications, covering distributed federation,
    distributed inference and acting on those results. It includes
    real-world examples of solutions that have been built using federated
    learning and discusses how to do federation across a wide variety of
    machine learning approaches"-- Provided by publisher.
Identifiers: LCCN 2021009936 | ISBN 9780367861575 (hbk)
Subjects: LCSH: Team learning approach in education. | Machine learning. |
    Artificial intelligence.
Classification: LCC LB1032 .V36 2021 | DDC 371.100285--dc23
LC record available at https://lccn.loc.gov/2021009936

ISBN: 978-0-367-86157-5 (hbk)
ISBN: 978-1-032-04935-9 (pbk)
ISBN: 978-1-003-02717-1 (ebk)

Typeset in Times New Roman
by Radiant Productions

Dedication

Dedicated to my wife Paridhi
without whose inspiration
this book would never have started

Preface

Artificial Intelligence (AI) holds a significant potential to transform the way many operations and processes are conducted in the real world. There have been many significant advances in the field of AI, primarily in the field of Machine Learning (ML), and that has led to many useful utilities, ranging from approaches that allow us to talk to machines, and for computers to use vision to perform a variety of functions, such as allowing access into secure premises. However, despite all the great advances made in the field of AI, it would be fair to say that we have just scratched the surface, and that a significant amount of impact is still to be made. While some businesses have made great use of AI to improve their business, many existing businesses, such as banks, retail chains, insurance companies, etc., have only seen limited usage of AI within their business.

The reason for AI not making stronger inroads into businesses is that there are some fundamental challenges that need to be overcome before AI techniques can be applied to solve real-world problems. One of the most basic problems that needs to be overcome is the access to data. AI techniques are a mechanism to capture the patterns that are inherent in the data, represent those patterns in a structure which we can call an AI model, and then use that AI model to improve the operation of a business. These two steps provide the main stages of an AI-based process, learning and inference.

When a large amount of data is available to create a good AI model, the application can be created in a very effective manner. However, when the collection of data is difficult or time-consuming, creating trained models on the data becomes difficult. Insufficient training data leads to poor models, and poor models are not useful for improving the operations of any business.

While it may seem that any existing business would have access to a large volume of data, which would render the problem of accessing training data moot, the reality is far from it. Regulated industries are constrained in how data can be shared among different parts of their business. The organization trying to improve the procedures using AI may not have access to large parts of data due to regulations that have been put in place for consumer privacy or security. Some businesses have grown through a sequence of mergers and acquisitions. In such businesses, the data may be split in many different locations with inconsistent formats. In some cases, moving large volumes of data from these locations can be very expensive, or take too much time. In other cases, businesses are structured into branches which may have limited network connectivity. In many of these situations, while the aggregate data volume may be large, collecting them into a single curated set may not be feasible.

In this book, we are examining this problem of creating AI models from data that is stored in many different locations, and can not be easily moved to a single location. Applying AI in this situation requires a federated learning approach. In this approach, local models are created at each site from the locally present data and these models are moved to a server which combines all local models into a single model. The approach for federation need not be limited to creation of models, but can also be used in the inference stage, when models are used for making decisions. In the federated inference approach, decisions can be shared across different sites in order to improve their quality.

Another issue preventing the adoption of AI is the gap between the theory of AI and adopting the theory into a real-world solution. AI/ML approaches have frequently been covered from the perspective of the statistician, and from the perspective of the data scientist or software developer who has to implement those functions. The application of AI to business problems must usually be done by a business analyst or senior technical architect who may not want to get bogged down in the details of the statistics or the complexities of software programming packages. To drive adoption into business, AI/ML has to be decoupled from the realm of statistics and programming and cast in terms which make sense to the business developer and the technical architect.

In order to focus more on the real-world system building aspects, this book has been written from the perspective of a non-statistician and a non-programmer. Instead of trying to prove complex mathematical properties, we have taken the approach of explaining the AI models and model-building exercise from the perspective of a business developer or a technical architect. In this book, we will minimize the use of complex mathematics and present the concepts around federated learning in intuitive business-friendly terms. Similarly, we will not go into the intricacies of software frameworks and packages that can be used to implement federated learning.

Our focus will be on use of AI in business settings as opposed to academic explorations around dealing with federation. Some of the hard challenges that a business needs to address do not make for good academic exploration. At the same time, some aspects of federated learning which are extremely interesting from a scientific exploration perspective may not be very interesting from a business perspective since simple work-arounds for the problem exist.

As an example of the former issue, handling data with different formats and inconsistencies takes up a significant amount of the time required for building federated business applications, but relies on approaches that do not lend themselves to exciting scientific publications. We have a chapter devoted to discussion of issues related to data inconsistencies. As an example of an academic problem that is not relevant to business, a significant number of publications have looked at issues of adversarial attacks in federated learning. While this provides many avenues for interesting research, businesses can often side-step the problem by signing business agreements and accepting the residual risk that exists after those business agreements have been signed.

The performance of AI-based approaches depends critically on the nature of data that is used to train and test the model. As a result, it is virtually impossible to make universally valid statements about the performance and effectiveness of AI algorithms. An AI-based model that works well on a set of publicly available images may have disastrous results on a set of private images taken to inspect the products of a manufacturing facility, and may be a good fit when applied to the task of network management by transforming network traffic to images. Any statements about performance of an AI approach are valid only in the context of the data that it was trained and tested upon. Therefore, we have avoided the temptation to show a lot of data driven results to compare different approaches in the book. Instead, we have taken the approach of identifying the qualitative reasons and the conditions under which some approaches should work better than the others.

The book consists of nine chapters which are structured as follows:

- Chapter 1 provides a non mathematical introduction to AI and Machine Learning. It describes the overall process for AI/ML model creation and for using those models in different business environments.

- Chapter 2 discusses the different scenarios within business environments which require the use of federated learning techniques. It discusses the motivation for federated learning, and also provides the distinction between consumer federated learning and business federated learning. The focus of this book is on the latter.

- Chapter 3 discusses the set of algorithms for federated learning that can be used when some simplifying assumptions can be made about the enterprise environment. These assumptions include the fact that data is available in same format at all the sites, data is uniformly distributed across all sites, every site is training the models at the same time, and that all sites have complete trust in each other. The subsequent chapters discuss the approaches that can be used when these assumptions are not satisfied.

- Chapter 4 discusses the issues that arise when the data is not in the same format at all the sites, and approaches that can be used to address the challenge of handling data with variable format and quality at different sites.

- Chapter 5 discusses the issues that arise when the data is distributed in a different manner across various sites, i.e. the data distribution is skewed and not identical. It discusses the approaches that can be used to address those situations.

- Chapter 6 discusses the issues that arise when the sites do not completely trust each other, and mechanisms they can put in place in order to work with each other despite not having complete trust in each other.

- Chapter 7 discusses approaches that can handle the joint learning of models when different sites can not easily collaborate in order to train models at the same time.

- Chapter 8 discusses approaches to share intelligence across sites that can not resolve the data mismatch or skew they have among themselves. These can arise in many situations where different sites are observing the same entities but collecting very different information about them.

- Chapter 9 shows how the different approaches and algorithms can be put together in some specific use-cases that arise in business. The use-cases are drawn from pilot engagements and discussions with clients in the business space.

The appendix of the book discusses some of the topics that are related to the subject of federated learning but not emphasized in this book. These include the topics of adversarial federated learning and software frameworks for federated learning.

We hope this book will be useful to the business and technical leader in understanding the power of federated learning and the approaches that can be applied to solve the challenges that arise in the use of AI in business.

Acknowledgements

Some of the research described in this book was sponsored in part by the U.S. Army Research Laboratory and the U.K. Ministry of Defence under Agreement Number W911NF-16-3-0001. The views and conclusions contained in this document are those of the author and should not be interpreted as representing the official policies, either expressed or implied, of the U.S. Army Research Laboratory, the U.S. Government, the U.K. Ministry of Defence or the U.K. Government. The U.S. and U.K. Governments are authorized to reproduce and distribute reprints for Government purposes notwithstanding any copyright notation hereon.

Some of the figures in this book has used icons from openclipart.org.

Contents

Chapter 1

Introduction to Artificial Intelligence

Artificial Intelligence (AI) has the potential to make a significant improvement to the effectiveness and efficiency of existing processes across a broad spectrum of business applications. However, like many popular terms, a precise definition of AI is not available, which leads to a wide divergence in the opinions of different members of the technical community as to what technologies are part of AI, and what should be considered outside the domain of AI. Intelligence itself has more than 70 definitions [1] and each text on AI has its own definition of the term. We do not want to decide what the broader definition of AI in the community ought to be, but it will be useful to define Artificial Intelligence in the context of this book.

The focus of this book is on the use of AI (and specifically the use of Federated Artificial Intelligence) to improve operations in real-world scenarios encountered in commercial businesses and military environments. Therefore, we will define AI in this specific context, first by using an abstract model representing the operation in any real-world scenario, describing how this model works with and without AI. This would provide a specific definition of what is considered to be in the scope of AI, and what is outside the scope of AI.

1.1 Business Operations Model

For the scope of this book, we define a real world scenario as making decisions in the course of a business operation through a software-based implementation. The exact nature of the business operation and decision would depend on the industry and on the specific organization within that industry. As an example, a bank may want to determine if a money transfer request made by a customer is legitimate or fraudulent. Another example would be for a telephone operator to determine if a phone call initiated by a caller should be allowed to go through or be blocked because the caller may be violating some norms.

Detecting whether a transaction initiated by a client is legitimate is an example of an operation required in many types of businesses, including but not limited to banks, insurance companies, telephone companies and retail stores. An initial check for fraudulent transactions can be done by checking existing records to ensure that the transaction initiator is an authorized client of the business, has presented proper credentials to authenticate themselves for the transactions and their account is in good standing. These checks are necessary but not sufficient. Credentials can be stolen, some customers may be deceived by a fraudster to make improper transactions from their account, and some criminals may have set up a legitimate account to conduct fraudulent transactions. Checking for fraudulent transactions requires additional tests for abnormal patterns of behavior. Determining what is normal and what is abnormal is an area where application of Artificial Intelligence techniques could be very useful.

Another decision that arises in many industries is determining which of their customers are high value, and which ones are low value. High value customers are those that may generate significant profit for the business, while low value customers may use a disproportionate amount of enterprise resources and generate loss for the enterprise. In general, any enterprise would prefer to give better service to customers that generate more profit. Restaurants would like to give special attention to customers who order expensive items on the menu, and may want to discourage patrons who just order a coffee and occupy a seat for long periods. Banks may want to give special incentives to clients who maintain a large amount of funds in their accounts and discourage clients who maintain very little balance with them.

However, it is not trivial to determine who the high value customers are. For example, a bank may want to give preferred status to the child of a customer who maintains a significant balance with the bank, even if the child currently is a student without significant funds in his or her individual account. The classification of customers into high or low value (or any number of tiers) needs to be done by examining the nature of their transactions with others, not just their the individual accounts.

Another type of business decision that businesses need to undertake arises during operations related to customer care and support. When customers call for help, the business needs to quickly determine the reason why the customer is calling, and route the call to the right system. The right system could be one of many automated systems that handle simple requests, or one of many human experts handling complicated requests. Determining which of the several human experts has the best combination of expertise and availability requires the software to make some complex decisions.

In addition to operations which deal with interactions between customers and a business, there are business operations which are inward facing and undertaken to ensure the smooth functioning of the enterprise. For example, when some type of message is seen on the network connecting different machines in an enterprise, it may require the configuration of the network to be modified in order to block a malfunctioning or malicious device on the network. Hospitals may want to examine the progress of their patients to determine which type of treatment is more effective, or whether a new type of illness or a new strain of a virus has started to manifest itself. Similarly, when the motors driving the air-conditioning in a hospital start to make a strange noise, its root cause needs to be diagnosed and the suitable repair process initiated. Several other operations of a similar nature are required within different businesses, each requiring one or more decisions to be undertaken on a regular basis.

There is a great deal of diversity in the type of business decisions and business operations that need to be undertaken. Some types of business operations are needed across many different types of businesses, while other types may only be needed in specific industries. Yet other types of business operations may only be needed in specific departments in a specific industry. Nevertheless, a lot of these business operations can be characterized by the simple model shown in Figure 1.1. The business operation consists of two steps, inference and action. The inference step converts the input to an output decision. The action step implements the decision that is made. One can view the inference step as the 'brain' of the business operation, while the 'action' step is the 'muscles' (arms/legs/limbs) that do what the brain commands them to.

In the examples discussed previously, the business operation of checking for fraud would result in a binary decision (is the transaction fraudulent or not). Depending on the decision, some action will be taken, e.g. the transaction may be allowed to go forward, a rejection notice is sent, or an automated audit check request is made.

The goal of AI is to improve the inference step of a business process. The inference step is implemented as a software module that takes the input and produces the decision as its output. Both the input and the output may be simple or complex. The input to the software module can be a single value, a file containing many values such as a spreadsheet, a feed from a sensor or a set of sensors, etc. The output similarly could be a single value, a set of multiple values, a file with a set of records, or a file produced in unstructured text. This output is the computer readable instance of the decision.

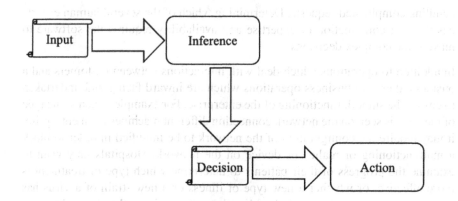

Figure 1.1: Idealized model of a business process.

More complex operations can be composed by combining many simple operations, with the output from one component operation feeding into another component operation. The same input can feed into multiple operations, resulting in a complex graph connecting several decisions and actions, and very complex business operation comprising many business decisions. However, we can easily understand the concept of AI by focusing on a single component, and its implementation as a software module.

There are many different ways to implement a business decision in software. Each of the different ways to do the conversion would map the input to an output, and all implementations would provide equivalent functionality. However, there may be a significant difference in the attributes of how the conversion happens, such as the skills of people who are required in order to implement the business process, the type of information that needs to be provided to them, or the speed at which the process can be performed. As an example, one approach to perform the conversion from input to output may involve hiring a team of software programmers, another approach may require having a team of data scientists, while a third approach may require having a team of people to enter data into a spreadsheet. Different approaches may also be more effective depending on the range of the values that the input to the decision process takes.

We will illustrate a few of these different approaches with a simplified version of a business operation. Let us consider the case when an applicant at a bank wants to apply for a loan. The bank needs to determine what type of risk should be associated with this loan application. In real life, this risk determination is based on many different factors, including the duration of the loan, the amount, the income of the applicant, the region of the country the person lives in, prevailing interest rates, etc. However, for the purpose of illustration, let us simplify the problem and assume that the bank makes the determination of risk on the basis of only two factors, the amount of the loan requested, and the annual income of the applicant. The bank would then determine the risk associated with the application as belonging to one of three categories—green, yellow and red, where green means the loan has very little to no risk, yellow means the loan has moderate risk, and red means the loan is very risky. In the inference module, the bank will determine the risk rating to be associated with the application and compute the risk given the two input values, while in the action module the bank software would determine how exactly to implement the decision that has been made.

The first approach that the bank can use to implement the inference module is to compile a table of risks associated with the different values of income and loan amounts. One possible example of this lookup table would be the one shown in Table 1.1.

Table 1.1: Lookup table approach for business decision.

Loan Amount	Annual Income	Risk
< 10,000	> 20,000	Green
10,001-50,000	>100,000	Green
> 50,000	< 100,000	Red
> 50,000	25,000-100,000	Yellow
> 50,000	> 200,000	Green
> 100,000	< 200,000	Red

Once the table is defined, the inference process consists of looking up the result value corresponding to any given input in the table.

Defining such a table may be a good approach in many cases. This allows the software to simply look up the information in that table, and determine the credit risk rating of a loan application. The operation would be to follow the approach shown in Figure 1.2. In order to maintain the table driving this software, the bank needs to hire a number of loan risk assessment personnel who can determine what the right entries in the table ought to be. It would also need a software engineer to write and maintain the code that draws the inference from the table.

While the table driven approach has the benefit of being easy to implement, the limitations of the approach should also be obvious. The table would need to be created and maintained based on the expert opinions of the risk assessment personnel. These opinions may be biased by their views as opposed to what the real inherent risk in the loan process may be. Over time, the table may not reflect the true risk based on the historical data about loan default, and may get out of sync with the reality on the ground. Also, over time, the table may grow large and may be difficult to maintain consistently. You may have noticed some points of

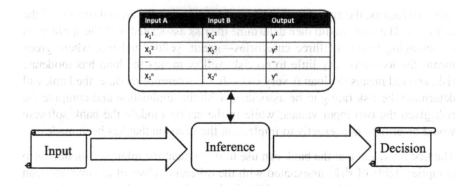

Input A	Input B	Output
x_1^1	x_2^1	y^1
x_1^2	x_2^2	y^2
–	–	–
x_1^n	x_2^n	y^n

Figure 1.2: Table-driven approach for business decision.

inconsistency in Table 1.1, in that some combinations of income and loan amount are overlapping, and the table does not cover some combinations. When there are multiple columns in the table, detecting those inconsistencies may be hard to do without software to analyze and identify them.

An alternative approach would be for the business to define a mathematical function that maps the different input values to a risk score. An example would be for the business to define a loan as green if the ratio of loan amount to income is less than 0.5, red if it is over 2.0, and yellow if the ratio is in between these two limits. Defining such a function would provide a compact representation of the information that is contained within the table. When the input is provided to the inference module, it computes the function and produces the desired output in the manner shown in Figure 1.3.

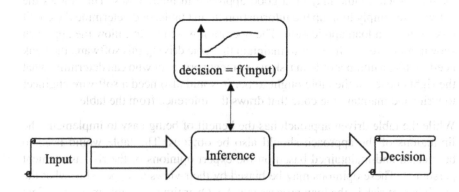

decision = f(input)

Figure 1.3: Function driven approach.

When such a function can be identified, this approach works very well. The module that implements the function, no matter how complex, can be created by means of a software implementation of the function.

The table and the function are alternative representations of different *'models'*. A model represents the relationship between the input and decision. There are many alternate representations of models to represent the relationship between the input and the output (decision). Some of the various representations of models are decision trees, support vector machines, neural networks, rule-sets, etc., and some of these are described in subsequent sections of this chapter. The model encodes the logic, using which a software module can translate an input into a decision to be used in the business process. The usage of a model to implement the inference step is shown in Figure 1.4. However, before the inference can use the model, the model has to be defined in some manner.

One possible approach is for an expert human, or a team of experts to define a model as shown in Figure 1.5. In effect, the model is capturing the intelligence embedded in the minds of the experts, and allows that to be implemented as a software module. Such an approach for defining a model is called symbolic AI. The model is defined in terms of some symbols which have meanings in the real world, and the symbols can be manipulated by the machine. A common example of symbolic AI is a rule-based system [2].

You may wonder how the expert derives their intelligence which they capture as a model. In most cases, expertise is gained via experience. A bank active in lending for a long period knows what type of people are more likely to default on their loans, and what type of people are more likely to pay off the loans based on their past history. This historical knowledge is captured into their models, e.g. life insurance companies compile extensive tables for the probability of deaths of people at different age groups in order to assess their risk.

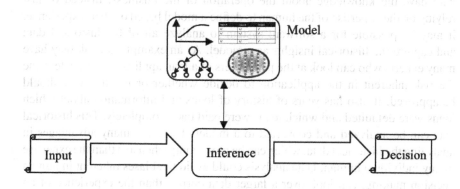

Figure 1.4: Model-based approach.

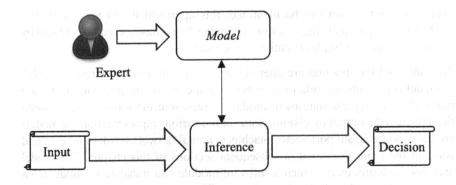

Figure 1.5: Symbolic learning approach.

Tools can be created to assist the human in defining the model more efficiently. As an example, tools can be created to help check for consistency within a model, check for common mistakes that a model can make, and allow validation of the model against some known ground truth. However, in symbolic AI, the human is ultimately responsible for transferring his or her knowledge into the model.

Symbolic AI approaches have been around for several decades, but they become difficult to use when models become very complex. In real-life situations, the logic of the model would typically have significantly more complexity than the simplistic example we have been shown above. There may be dozens of inputs needed for assessing the risk of a loan application, e.g. the availability of a collateral, the past payment history and credit risk score of the applicant, etc. Defining a symbolic model with a large number of inputs is a non-trivial task by any human.

In most businesses, there is significant historical data as well as a set of experts who have the knowledge about the operation of the business. Instead of just relying on the expertise of the human to define a model based on their experience, it may be possible for a software system to analyze all of the historical data and capture the historical insights as a model. As an example, a bank may have many experts who can look at the parameters of a loan application and determine the risk inherent in the application to decide whether or not the loan should be approved. It also has years of history of loans and information about which loans were defaulted and which loans were paid back completely. This historical data can be analyzed and converted to a model. There are many advantages in analyzing the historical data, which can be more comprehensive than the expertise of any individual human. Data analysis could avoid the biases inherent in human decision making, can look over a larger data corpus than the experience of an individual human, and can discover relationships and patterns that may not be obvious.

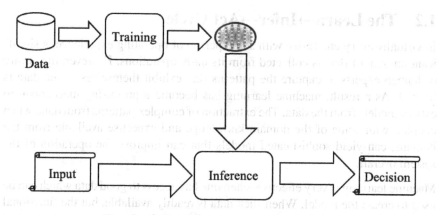

Figure 1.6: Machine learning approach.

The mining of historical information using automated computer programs to create a model is called machine learning. The step of creating the model is usually called training or learning, and the model can then be used to perform the inference task. This process involved in machine learning is shown in Figure 1.6. A popular type of model that is often created is a neural network (See Section 1.5.5), so machine learning techniques are also referred to as neural learning.

Both the human expert and the data mining process have valuable and complementary roles to play in the development of the model, discussed in more detail in Section 1.4. In real-life, hybrid approaches which rely on a combination of both human expertise (Symbolic AI approaches) as well as machine learning are more likely to be useful in improving the solution of a business problem. One specific type of hybrid approach would be the case where humans write down their expertise in documents or reports. Computer software capable of processing natural language processing can then be used to analyze those documents and create the model. The data for machine learning in this case is the captured knowledge of humans which is represented in an indirect format, such as text. Sometimes, the text-based knowledge may be captured in different documents that may be present in a distributed environment like the Internet. A computer software may crawl the Internet, collect the relevant documents, convert them into a model and augment that model using analysis of historical data. There are many different ways of creating AI models, and in practice all useful models would use a hybrid approach of mixing symbolic AI with some machine learning approach.

For the purpose of this book, we define an AI-based method as an approach for implementing business processes which is driven by the use of a model that in itself is derived using software-driven analysis of data.

1.2 The Learn→Infer→Act Cycle

In virtually every enterprise with some period of operating experience, a significant amount of data is collected from its daily operations. However, the ability of human experts to capture the patterns that exhibit themselves in the data is limited. As a result, machine learning has become a promising mechanism to extract models from the data. The extraction of complex patterns from data, when coupled with some of the domain knowledge and expertise available from the humans, can yield sophisticated models that can improve the operation of the system overall.

Machine learning is very effective when one has access to good data which can be used to create the model. When such data is readily available, but the functional relationship among different components of data is hard to determine by human inspection, machine learning would be the preferred approach. When the training data is hard to collect, and the patterns required for inference can be easily defined by an expert, symbolic AI would be the preferred approach.

In practice, every model is imperfect, and can be made more accurate over time. The model may be learned, then used for inference, and the results of the inference used to take specific actions. The actions in themselves would reflect them into some operational metrics of the business process. The reaction of clients, business associates and employees on specific actions may lead to changes in the relationships between the input and the output. Over time, the system would get additional data which may require the model to be retrained. As a result, it is better to think of an AI-based application as having a life-cycle in which the learning, inference and action phases inform each other. This life-cycle model introduced initially for military operations [3] but valid in most business contexts is shown in Figure 1.7.

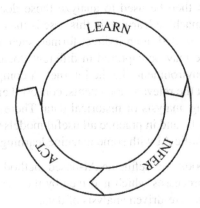

Figure 1.7: Life cycle for AI-enabled business process.

Learn Phase. In the Learn phase, AI models are created afresh or adapted from existing models. Using software packages to analyze data, the process creates a model that is capable of converting input to output. The conversion is done so that it is faithful to the relationships contained in the training data. This training process can be very compute-intensive since it is analyzing a large volume of data. After the model is built, the learning process will also include additional steps of testing the accuracy of the data on validation data. For proper assessment, the validation data should not be a subset of the training data.

Infer Phase. In the Infer phase, the input data is used to produce an output decision for the business process. Inference is the capability of the model produced in the learn phase to be applied to a new data point to produce a decision. The inference step may consist of a pipeline of activities, where the initial stages of the pipeline would consist of converting the input provided to the system to the set of inputs that the AI model is expecting. In general, the input data that is generated in the course of business may not correspond exactly to the assumptions about the input made by the AI model, and feature conversion needs to be done explicitly. The latter stages of the pipeline perform the actual step of invoking the AI model.

Act Phase. In the Act phase, output of Infer phase is used to take an action or implement the decision. The type of action/decision can be varied, and may involve input from humans, other software, or other machines.

In principle, the same model can be defined by the expert as well as learned by the machine during the Learn phase. However, some models are easier for humans to define, e.g. a set of rules or a set of tables with different columns. Some other types of models, which require many complex weights and complicated branching decisions, would be difficult for humans to define, and are better generated using machine learning techniques.

1.3 Process of Creating an AI Model

Creating AI/ML models requires a systematic approach to perform a sequence of activities. The process for business processes, which is an adaptation of the process for military multi-domain operations [3], includes:

■ **Identify AI/ML tasks** that can be improved with the use of AI/ML techniques. Not all tasks may be suitable for such improvement. In general, tasks which require very simple models may be implemented by means of software programs, or with approaches like a table lookup. However, for tasks whose models can not be captured easily, and where adequate good quality data for training is available or can be easily obtained, the use of ML techniques may be more appropriate.

- **Design AI/ML to meet the goals** of the task under expected constraints in its intended operational setting. This design would require the choice of the model, decisions as to whether a symbolic or machine learning approach is to be used, and what criteria ought to be satisfied for the AI/ML task to use for the model.

- **Obtain and curate relevant training/validation data**, when the task is based on ML. Training data is used to create a model, while the validation data is used to assess the performance of the model. Ideally, the training data and validation data would be independent. However, since procurement of the data is often the most time-consuming activity in the entire AI/ML process, the training data set is frequently divided into two different sets, one used for training and one used for validation.

- **Train models under a variety of conditions**, including using dirty, dynamic, and deceptive data that will be present in the operational environment. It is important that a model training cover as many situations as are likely to be encountered in the operational environment. Generally, models would not perform well if they are asked to predict information in an environment which is very different from the one they were trained on.

- **Validate AI/ML in realistic conditions** in order to ascertain that the model is suitable for purpose. The validation is done using the validation data, and it is assumed that the validation data reflects the operational environment.

- **Analyze the AI** to understand and predict its performance and **assess** its safety, security, and robustness. It is at this stage of the process that symbolic methods have an advantage over ML-based methods, since the analysis of symbolic models is usually easier than that of machine learning based models.

- **Determine the allowable use cases** for the AI/ML-enabled task to constrain how it can be used (including the goal of ensuring it is within ethical guidelines), determine allowable sharing arrangements, how and when learned models can be adapted or updated, and where and when can models be trained, inferences can be made, and action be taken

A key element of this process is to assess what parts of the business process are appropriate to be AI-enabled, what level of autonomy is desired, and what the role of the human in the overall AI-driven business process would be.

1.4   The Role of Humans in AI

One of the concerns often seen in popular press about AI is that AI techniques can replace humans, which can cause tremendous social upheaval. Therefore, it is important to discuss and assess the role that humans play within the framework of AI-enabled tasks for business operations. Considering the Learn→Infer→Act as the basic loop of the AI process, human involvement is required at different stages of the cycle.

1.4.1 *Human Role in Learn Phase*

During the learning phase, the system needs to learn a model from the training data, or has a human provide its learned knowledge into the model. The role of the human in defining the model is obvious for symbolic AI approaches, where the human is providing its knowledge to the system. The human input is also essential in ML-based approaches.

A key challenge in machine learning for model creation is the sensitivity of the learned model to available training data. The quality of the model is dependent on the quality of the training data that is provided. Significant human involvement is required to process the data and ensure that is it of high enough quality to train a good model. This human effort and tedium can be reduced by tools that assist in the collection and analysis of the collected data, but the data collection step is something that is hard to automate with the current state of technology.

Another challenge associated with the model building phase is that the training algorithm can over-fit the model to the data. The data may have patterns within themselves that the model could pick up, which may lead to incorrect and undesirable results. Human involvement is required in order to ensure that the learning model is not picking up these undesirable patterns. As an example, a model to recognize fruits may pick up the red color as the distinguishing property of apples because all of the images used in the training used images of a red apple instead of its shape. Human intervention is required to ensure that the patterns learned by the model from training data are proper, and not picking up spurious patterns.

With the current state of technology, AI in business process is best used as an assistant to augment the model building process that a human being would undertake. The computer-based data-driven analysis can identify many interesting patterns which would be hard for a human expert to extract from the data. However, the human expert would be able to determine if some of those patterns are spurious and which of the patterns are useful.

In order to assess the usefulness of models, the model must be understandable by a human. Some types of models are easier to understand than others, and interpretability of the model is an important aspect for vetting by humans and for use in business applications. As a result, some models that may be very good at performing the functions, but not be understandable or explainable, may be less desirable from a business perspective compared to a model that is understandable and whose outputs can be explained, even if this model is under-performing on an absolute scale in its accuracy or performance metrics.

1.4.2 *Human Role in Infer and Act Phase*

In the Infer stage of the Learn→Infer→Act cycle, the inference step can be completely automated. It can be argued that humans need not have any role in the inference process, which produces a decision. The act stage of the cycle, however, is the step where human involvement becomes desirable depending on the business task which is being performed.

In many business processes, the output of the inference stage is a decision which is advisory in nature to a human. The human has to decide on the right action to take based on the output of the decision process, and is an integral part of the overall cycle. This model of Human in the Loop (HIL) is the norm in many AI enabled business processes.

An alternative model for involving humans during inference is the Human Over the Loop (HOL) model [4]. In this model, the inference and resulting action phase is automated, but the human has a supervisory role over the inference and action phases. The HOL model can be seen in autonomous self-driving cars with a human behind the wheel who can take over the driving task at any time, or in the auto-pilots for airplanes where the pilot can take over as needed.

The selection of the approach used for actions, whether it should be completely manual, Human in the Loop, Human Over the Loop, or completely automated depends on the consequences of the wrong action being taken and what is the relative risk of choosing a wrong action between a human and an automated process.

For actions where humans are more likely to make a mistake, and the impact of a wrong action is not significant, complete automation may be the right approach to follow. As an example, for repetitive tasks, like sending an email out in a fixed template, humans are more likely to make a mistake than automated systems, and the impact of a wrong email is not significant in most cases. That task is better off automated.

In a task where a machine is more likely to make a mistake, e.g. in a complex decision-making process for handling an emotionally upset customer, and the impact of losing a customer is significant for the business because it is a high value customer, complete human handling may be the best choice. In between these two extremes, human in the loop and human on the loop mechanisms can be followed to provide the best trade-off between the impact and risks.

1.5 AI Model Representation

An AI model is in effect an alternative way to represent a function which takes several inputs and produces an output decision. There are many different ways to represent such a function. We have already examined a couple of different techniques to represent such functions, including the representation as a table and the representation as a mathematical function. In this section, we look at the various other alternative representations of AI models.

Given the wide diversity of different AI models, this section does not consider all of the possible variations of AI models, but only considers a few common types that have wide applicability within business operations.

1.5.1 *Functional Representation*

The most general representation of the AI model is as a function. For this section, we can assume that a function consists of several input values, let us say N input values $x_1, x_2, \ldots x_N$ and has an output y. The generic function that is being represented here is a function $y = f(x_1, x_2, \ldots x_N)$.

When this function has to be represented and fed into a computer software, it can be be represented as several parameters which provide a suitable approach to calculate the function as needed. The function may be modeled in terms of different assumptions as to how the output depends on the inputs. In some instances, one could assume that there is a linear relationship between the input features and the output features, i.e. the relationship can be captured by means of the equation:

$$y = \alpha_0 + \sum_{i=1}^{i=N} \alpha_i x_i$$

The AI model would in this case consist of the $N + 1$ parameters of $alpha_0, alpha_1, \ldots alpha_N$. A common method to find an AI model for a linear relationship is to use the techniques for linear regression [5].

In real-life, not all relationships are linear. Relationships can also be captured as non-linear equations between different powers of the input variables, or as a result of the product or division of the different input values. A set of parameters can represent any complex combinations of the relationship between the inputs and the outputs. As an example, a relationship between the same inputs and output where each input could have a power of M max can be represented in the following relationship.

$$y = \beta_0 + \sum_{j=1}^{j=M} \beta_{1,j} x_1{}^j + \sum_{j=1}^{j=M} \beta_{2,j} x_2{}^j \ldots + \sum_{j=1}^{j=M} \beta_{N,j} x_N{}^j$$

This can be represented as a matrix of size $Nx(M + 1)$ by breaking the constant term β_0 into $M + 1$ constants that sum up to it.

Another common approach to capture non-linear relationships is to take the logs or exponential of the output or input variables. Then, a regression can be plotted between the logarithm of the output and the inputs or other combinations thereof. The approach for logistic regression [6] is used in many areas and provides a way to capture non-linear relationships.

There are many other ways to represent a function using a set of parameters in the model, as combinations of other functions, or as sums, products of inputs, or as exponential or log representations. Usually, the mathematical functional representations are good to use for models that represent physical phenomenon, e.g. when one is trying to understand the operation of mechanical parts like car engines performance, conditions of mechanical equipment on the factory floor, or make sense out of the information produced by a sensor in Internet of Things (IoT) [7] application scenarios.

Functional representation can be useful in many real world situations, ranging from determining the right control and configuration settings for physical control of equipment and machinery, making financial decisions such as estimating the default risk associated with a loan, predicting the value of a house, estimating the configuration that will maximize the performance of a data center or network, etc. It is also the most general representation of the relationship between the inputs and the output, and can be used to approximate any other model. However, for some specific cases, other types of model representations may be a better fit, specially when some of the input or output values are not numeric.

1.5.2 Decision Tables

While functional representations are a good way to have a general representation of any phenomenon and capture the relationships, many business considerations would not be modeled by relationships that may be easy to represent in a mathematical formulation. The table of risk associated with business loans shown in Table 1.1 has three categories of red, yellow and green as its output. One can still calculate a mathematical function and associate the three colors for different ranges of the output of the function, but that function is not likely to have a very simple representation. Its graphs would be changing abruptly, and have a lot of discontinuity. Business phenomena that do not rely closely on physical devices but on human experience, social relationships and gut feelings are hard to model as simple smooth mathematical functions.

Inputs and outputs could be either numeric values or be categorical values. Numeric values are numbers which would typically have a lower and upper bound depending on the business context, with no bound being a special case. Categorical values are ones which take value as one of different discrete categories, e.g. names. The output in Table 1.1 which consists of three color names is categorical.

Instead, a better way to model these functions would be to list them out in a tabular format like the one shown in Table 1.1. This representation of an AI model is a decision table [8]. When used to represent the function with inputs $x_1, x_2, \ldots x_N$ and output y, the decision table will have upto $2N + 1$ columns. Each of the input variables that is numeric would have two columns, one marking the lower value of the input variable, and the other marking the maximum variable of that input variable. Each of the input variables that is categorical would have one column with the category of the value.

Each row in the table provides the value of the output if an input data falls within the conditions defined by the bounds and values of that row. The decision table can be seen as a compact representation of the training data, which is also a tabular representation with a column for every input and an additional column for an output. The decision table summarizes the training data set which has a much larger number of rows into a more compact representation. During the compaction process, there may also be conflicting information in the training data in the sense that two data points with the identical or very similar input may produce very different output. The training process would resolve those conflicts in order to get a self-consistent system to map inputs into outputs.

When output decisions are categorical, decision tables can provide a good representation for defining an AI model.

1.5.3 *Decision Trees*

The decision tree [9] consists of a tree-like structure consisting of several nodes. A tree-like structure means that there is a starting or root node, and each node may have some children. Nodes which do not have any children are called leaf nodes. Each node in the tree contains a test on one input variable, and the output of the test dictates which of the children of that node will be selected for a subsequent test. At each leaf node, there is a predicted value of the output.

A simple example is shown in Figure 1.8. Each node in the tree has two children. To determine the risk of a loan into red, green or yellow. At the root node, which is the left-most node in the diagram, the test checks if the loan amount is less than a threshold (100K in the example). If the test fails, the loan is marked as red. If the test succeeds, a second test is conducted which checks if the salary of the applicant is less than 50K/year. If the test fails (i.e. the salary is over 50K/year), the loan is marked as green. If the test succeeds, the algorithm runs another test which checks if the loan is less than 10K. If the test succeeds, the loan is marked green. If the test fails, the loan is marked yellow.

Real-world decision trees take several dozens of parameters as input, and are much more complex than the simple example shown above. However, they work

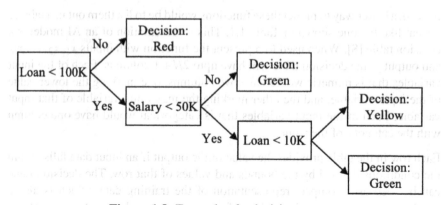

Figure 1.8: Example of a decision tree.

on the principle of conducting a test at each node and following different branches of the tree depending on the outcome of the test.

A decision tree can be viewed as a compact and efficient representation of a decision table. If a decision tree is binary, i.e. each node has two children, a tree which performs N tests in one pass of the tree can represent the model which requires 2^N rows in a decision table.

The training of a decision tree requires the input training data to be scanned in order to find the most efficient set of tests that can lead to a rapid decision making. There are a variety of efficient algorithms that try to find the best representation based on comparing the statistical properties of the data set, and the impact of adding a node in the decision process.

1.5.4   *Rule Sets*

A rule set [10] is another representation of an AI model, which consists of a number of rules. In each rule, there are a set of conditions involving the different input variables, and an output value if the conditions are true. The conditions are a logical relationship that can hold between one or more input variables. In contrast to the decision trees where each row contains something for each of the input variables, a rule can only contain a subset of the input variables, and ignore the others.

Some rules can also predict another input variable instead of directly predicting the output. These types of rules can be used to predict a new input value which can then feed into another rule to predict an output in another step. Rules can be chained as well.

If all rules in a rule set predict the output directly, then they can be converted easily to an equivalent decision table by converting each rule to a row of the decision table. Any path in a decision tree can also be viewed as a single rule which consists of all the checks along the path performed in sequence.

Rule sets are one of the earliest mechanisms to represent an AI model. They can be written by hand by a human expert or they can be determined by analyzing an available training data. They have been used in many different types of systems, and have proven their value in many business systems.

1.5.5 Neural Networks

A neural network [11, 12] is a representation of AI model which is inspired by how the human nervous system works. A neural network consists of multiple nodes interconnected in a network where each node (called a neuron) has multiple inputs and one output. A function defines how the inputs are mapped into the output. The nodes are usually arranged in layers, each layer consisting of several nodes, and the output of one layer feeding into the inputs of the next layer. The first layer takes in the different input variables $x_1, x_2, \ldots x_N$ and the final layer predicts the output value y. The function of each node is defined by a set of parameters generally referred to as weights. As an example, a neuron may take several binary valued inputs and produce a zero if the sum of inputs exceeds a threshold, and one otherwise. The threshold would be the weight for the neuron. The collection of all the weights at all the neurons in different layers can lead to a neural network being able to model almost any type of relationship between the input and output variables.

The flexibility of neural networks is further expanded by the fact that there are several functions that can be used with each neuron, different layers can use different types of functions, and a very rich network topology can be generated from them. As a result, neural networks can represent any arbitrary function and over time several efficient algorithms for training them have been developed. Recently, deep neural networks which consist of several layers of neurons connected together have proven very successful in a variety of applications, including image recognition, speech to text conversion and natural language processing. The increase in affordable computation power over the years and access to large amounts of training data has led to an explosion in the application of neural networks in many different domains.

Neural networks come in many forms, and depending on their interconnection topology, number of layers and type of functions encoded within neurons at different layers are defined into one of many different categories. These include feed forward neural networks, recurrent neural networks, convolutional neural networks, long-term short-term memory neural networks, Boltzmann machines, Hopfield networks, etc. Each variant has features which allow them to address a particular class of problems really well. As an example, convolutional neural networks work well for understanding the content of images and classifying them while long-term short-term memory or LSTM networks work well for modeling time-series type of data. We would not go into the details of specific types of neural networks. The approaches for federation covered in this book would be applicable for all types of neural networks.

The advances in deep neural network capabilities is one of the primary drivers for the recent interest in applications of AI to many different business problems.

1.5.6 *Matrix Transform-based Representation*

Some AI models can be represented as matrices, i.e. an array of NxM numbers which are multiplied to the input variables $x_1, x_2, \ldots x_N$ and it results in another set of transformed variables $v_1, v_2, \ldots v_M$. Another set of operations, e.g. a weighted sum on the output variables can then predict the output y. The mapping of the transformed variables to output can also be done by another AI model, e.g. a decision tree or a neural network, effectively coupling two AI models together.

A common application of the matrix transform models is to reduce the number of features into a smaller set of variables that have less cross-dependency among each other. This transformation is usually done using principal component analysis methods [13] for input variables that are numeric and using multiple correspondence analysis [14] for input variables that are not numeric but take some discrete values. Both of these approaches result in a matrix format for representation of the model. Matrix representations can then be used as a pre-processing step before learning other models.

1.5.7 *Distance-based Models*

Matrix transform-based methods can also be viewed as converting the input variables into a transformed space of many dimensions. The transformed space can then be used as a representation where distances between the transformed spaces can be calculated. A common approach to calculate distance between two points $v_1, v_2, \ldots v_M$ and $u_1, u_2, \ldots u_M$ is calculated using the so-called dot product among the different vectors, i.e. the distance between two points is calculated as $v_1.u_1 + v_2.u_2 + \ldots v_M.u_M$. The transformation to a space where distance is defined can be done using matrix transform or, more generally, using a neural network. A common approach to map inputs to a vector representation using a neural network is through an auto-encoder, a neural network that consists of an encoder part that maps the original input to the vector, and a decoder that maps the vector back to the original format. The comparison of the reconstructed input can be used as a measure of goodness to train the auto-encoder.

If a distance metric can be defined among different points, then methods based on distances can be used among different points. One specific approach used in many approaches is to find groups of points that are close to each other. This can allow determining inputs that are close together, and address functions like clustering 1.8.2, anomaly detection 1.8.3 and filtering 1.8.5, which can in many cases be done without knowing what the output y ought to be.

Distance-based methods can also be used to define surfaces in the transformed space which can provide a mechanism to predict categorical values of the output y. A popular algorithm to calculate these mechanisms includes the concept of support vector machines(SVM) [15], which finds surfaces in the transformed space to best separate different types of output space using efficient techniques to support both linear and non-linear relationships between the inputs and the output.

1.5.8 *Finite State Machine Models*

A state machine based model [16] defines a system to be in multiple states, each of the states defined by a combination of inputs that are being monitored. The system could be defined to be in some K states $s_1, s_2 \ldots s_K$, where each state is defined by some combination of input variables $x_1, x_2, \ldots x_N$. Each state is associated with a specific action that should be taken when the system is in that state, and the action corresponds to the decision y. Each of the states is also associated with a set of rules which specify when the system ought to change state.

The state machine can be defined to be either deterministic or probabilistic. In a deterministic state machine, deterministic rules dictate the next state to be taken. In a probabilistic state machine, the choice of next stage is random, driven by some probability distribution. The sum of all probabilities for next state needs to be 1.

State machine-based models have been used successfully in several practical applications, including robotic controls, self-driving cars, computer games or test case generation for software testing. Training techniques for state machine-based models include approaches called automaton learning and reinforcement learning.

1.5.9 *Equivalence of AI Models*

While we have given a very brief overview of some popular types of AI models, there are several other types of AI models. These include stochastic models which include Markov chains [17], Hidden Markov Models [18], Petri nets [19], etc. Within each type of AI model, there are several variations themselves. The same type of model, when used in different application domains, can be referred with different names.

While there are many different variations on how the model can be represented, all models are equivalent and one can convert one type of model into another type of model. The effectiveness of the model training process, the access to existing software packages that produce specific type of models, and the background of business personnel creating the models are key determinants in selecting what type of model to use.

The equivalence does not mean that all AI models would work equally well on the same data sets. The performance of each model depends on many different aspects, including the specific hyper-parameters chosen to define the model architecture. Such hyper-parameters include the number of nodes in a decision tree, or the number of layers and neurons in each layer selected for a neural network, or the number of dimensions of transformed space selected in a matrix based method, etc. The time taken to train the model, the ability to capture the models to capture the different patterns present in the data, and the ability to understand and explain how the model is working are all factors which will be different for different models. However, any model can be used as part of the inference process, and the definition of the the best model for any task at hand would be highly dependent on the specifics of the task.

Because of their flexibility in representing different types of functions, we will focus within this book on the functional representation of the model and neural networks as the exemplar AI models that need to be federated. However, the same concepts and approaches for federation would apply to other types of AI models.

1.6 Model Learning Approaches

There are many different approaches for training AI models. Each approach results in the creation of an AI model. In some cases, the representation of AI model is closely associated with the type of learning approach. Within each approach, there are many different algorithms that can be used to train the model, with each algorithm having its own set of advantages and disadvantages.

The approach for learning a model is characterized by its training data and the attributes of the training data that is available. Some of the common approaches of learning models are as follows:

- **Supervised Learning**: When training data consists of input data and corresponding output (label), one can use a supervised learning method to create a model. A label may identify a discrete class, a continuous computed value, a state, etc. Supervised learning is considered task-driven since its usage is dependent on the type of task being done. Some tasks which can successfully use this style are classification (e.g. images) and regression (e.g. prediction of target tracking).

■ **Unsupervised Learning**: When training data consists of input information without a label, one can use an unsupervised learning method to create a model. Example problems include clustering, anomaly detection, association rule learning, dimensionality reduction (identifying the most significant features of the data). Unsupervised learning is data-driven.

■ **Semi-supervised Learning**: When part of training data has a label, part is unlabeled, one can use semi-supervised learning schemes. This focuses on combining labeled and unlabeled data, usually using heuristics to label the unlabeled data. Unsupervised transfer learning can be used when the task is the same but the operational domain is different from the trained domain. Example problems include: speech analysis, knowledge organization, classification, regression. Semi-supervised learning is hybrid task-data driven.

■ **Transfer Learning** : When training data is available in one domain, but only limited training data is available in a related domain, transfer learning can be used. This involves adapting a previously trained model to a new related target problem. This extends the unsupervised transfer learning to the cases where the trained and operational tasks are different (induction transfer learning) or when the tasks are the same but the operational domain is different than the trained domain (transductive training). Example problems include: document classification and learning from simulations. Transfer learning is for new domains.

■ **Reinforcement Learning** : When the output consists of a goal for the system, one can use reinforcement learning to try to create a model that moves towards that goal. This focuses on learning from experience, usually by giving it a reward, and balances exploration and exploitation of current knowledge. Reinforcement learning can be considered a way to react to the environment to achieve a goal. Example problems include: robotic mobility, games, predictive maintenance. Reinforcement learning is goal-driven.

The summary of machine learning approaches and when to apply them based on the characteristics of training data that is available is shown in Table 1.2.

1.7 Spatial and Temporal Aspects of Learning

The learning process of the learn-infer-act cycle can be deployed in a few distinct ways. In one mode, the learning can be static, things are learned once usually in a controlled or enterprise setting, and assumed to be static or quasi-static. In the

Table 1.2: Training data and machine learning approach.

Training Data Characteristics	Learning Approach
Contains output labels	Supervised Learning
Does not contain output labels	Unsupervised Learning
Limited data with output labels and large data without labels	Semi-supervised Learning
Limited data but large data available in a related area	Transfer Learning
Set of actions with their impact on system	Reinforcement Learning

other mode, learning needs to be continuous and the models updated on a constant basis in a deployed system. The common ways for deployment of learning are:

■ **Offline or Batch Learning**. In this case, the AI models are created in an offline setting. If symbolic methods are used, the human input is collected in an offline method. If machine learning is used, training data is collected offline, the model may be trained on a portion of available training data, and tested for performance on another portion of that data. Subsequently, the trained models are frozen for deployment where inferencing uses the model that has been trained before. This approach is best used in less dynamic/uncertain situations as it has better verification and known performance but does not adapt to new situations or mismatches between the trained model and the environment. The model, which is static and immutable during deployment, can be updated after some time with a newer model that is trained in an offline manner.

■ **Online or Continuous Learning**. Train/test offline, but the learning algorithm is deployed to adapt the model with new input data. Thus, it continuously learns by incrementally adapting as data becomes available after deployment into its operational context. This approach is best used in dynamic/uncertain situations where the risks of adaptation are out-weighed by the need to react to unknown or changing circumstances.

From a spatial perspective, learning can happens in a central location or many different locations. The major options for learning are as follows:

■ **Centralized**: Centralized Learning builds its model in one location where all of the training and validation data resides. After the AI model is validated and verified, it is frozen prior to its deployment. This form of learning is particularly useful with large data-sets that can exploit high performance computing and high-speed networks, but is not appropriate when there are training data sharing constraints.

■ **Distributed**: Distributed Learning extends the centralized approach with distributed processing to overcome computational and/or storage con-straints, often using parallel processing techniques. This is frequently used in enterprise environments with fast local area networks that can distribute data to multiple processors.

■ **Federated**: Federated Learning is collaborative training where training data is not exchanged. In Federated learning, the training is distributed without exchanging training data in contrast to Distributed ML where training data is moved to be processed by multiple computers. Federated ML is used to overcome constraints on training data sharing (policy, security, sharing constraints) and/or insufficient network capacity.

From the same spatial perspective, the major options for inference are as follows:

■ **Centralized**: Centralized inference conducts its inference in one location where the model for inference is available. Usually, it is the same location used for the training task. If the location has many machines it is possible that the server used for inference is not the same as the server used for training the model, but the difference is usually not visible to any system outside the location.

■ **Edge**: Edge inference conducts its inference in a location close to the location which is generating the data. The inference site is usually a system located at some type of boundary, e.g. the system that connects a remote location to the central location which is the edge. In edge inference, the model is moved from the training site to the system at the edge and used for inference. Many edge sites may be concurrently using their systems for their local sites.

■ **Federated**: Federated inference is a variation on edge inference where different edge sites can collaborate together to improve their decision process.

The focus of this book is on federated AI, which covers both federated learning and federated inference. Federated AI is an important requirement in many real world situations. In the next chapter, we look at some of the common scenarios where we may need to implement federated machine learning solutions.

1.8   AI Enabled Functions

AI-based techniques can be used for many different business operations. Some of these business operations are specific to the industry or the enterprise, while some of the business operations can be viewed as applicable to a broad set of enterprises across several industries. Many of these business operations are based on some common functions.

We can divide these functions into three broad categories shown into a layered architecture as shown in Figure 1.9.

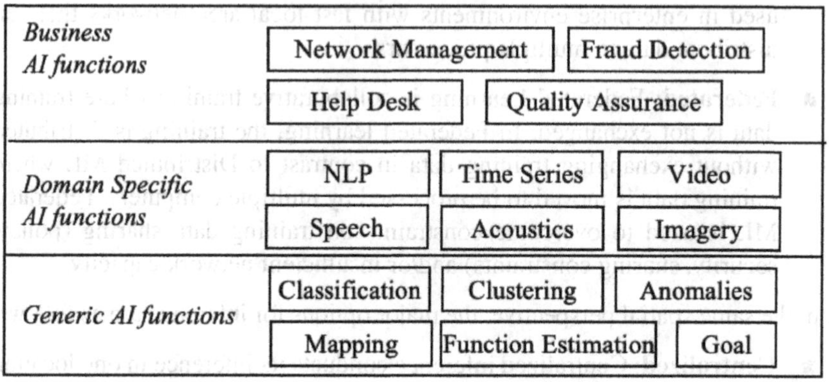

Business AI functions	Network Management		Fraud Detection	
	Help Desk		Quality Assurance	
Domain Specific AI functions	NLP	Time Series		Video
	Speech	Acoustics		Imagery
Generic AI functions	Classification	Clustering		Anomalies
	Mapping	Function Estimation		Goal

Figure 1.9: Architecture of AI-enabled functions.

■ **General AI Functions**: These are functions that provide a general capability based on the use of AI/ML techniques.

■ **data-type based Functions**: These functions apply AI/ML to a specific type of data. These functions may reuse some of the general functions, but frequently customize the general functions in a way that is optimized for handling a specific type of data.

■ **Business Functions**: The business functions build upon the general functions as well as data-type based functions to solve a specific business problem. The business functions can reuse algorithms and approaches from both the general functions and the data-type based functions.

Note that each of these functions are sufficiently complex within themselves to warrant an entire book. Our goal in this section is to provide the reader with a broad overview using a consistent terminology so that the main concept in the book, performing model learning and inference across different sites, can be explained in detail subsequently. For explaining the operation of federated machine learning, we would focus on the general AI functions, as opposed to data-type based functions or business functions.

The data-type based functions are AI/ML-based functions which support one given type of input data. These functions include the processing of time-series data, graph data, speech sounds, non-speech sounds, images, videos, text documents and network traffic processing. In each of the data-type processing, some of the capabilities may be derived from the specific properties of the input data-type. These can then be combined with AI/ML-based techniques to solve specific problems.

Business functions solve a specific problem that may arise in a given industry. Such problems build upon the general functions as well as data-type functions, and may use AI/ML-based capabilities only for specific activities. Some of these business functions may be used across multiple industries. Examples of such functions include loan risk assessment, and fraudulent transaction detection in finance industry, production quality optimization in manufacturing industry, and Intelligence, Surveillance and Reconnaissance (ISR) in military environments. Some functions that can be applied across multiple industries include detecting network intrusion attacks, handling customer care requests, matching applicant resumes to open job postings, and improving network and system management for IT installations.

There are many domain specific applications and many business functions of AI, each of which are too numerous to describe. We will only describe some of the common types of generic AI functions further in the various subsections below:

1.8.1 *Classification*

Classification is an application of AI techniques in which the output decision identifies the input as belonging to one among several classes that are usually pre-defined. Classification is required in many business applications, such as marking a loan application as high-risk or low risk, determining if the image of a product corresponds to a good product or a defective product, determining if the sound emanating from a device is a squeak, a clank or a screech, examining a radiology image to determine if it shows a healthy condition or one of a set number of diseases, examining if a piece of software downloaded from the Internet is one of many types of known viruses, etc.

For classification, the training data consists of several examples of inputs belonging to each class. During the training processes, the system examines the different inputs to extract the patterns that characterize the input belonging to each type of the class, and stores that as an AI model. Many types of AI models can be used to capture the patterns that define each class, ranging from neural networks and statistical separation to decision trees and rules. Some of the models can be defined by a human being instead of being learnt from a training data.

The effectiveness of classification is measured by means of its ability to accurately classify the results of the machine learning approach with what a human being would do. During the algorithm testing period, this can be done by checking the performance of the algorithm on a test set. If a training data set is available, it can be split into a training set and a test set, with the latter being used to check for measure of accuracy.

1.8.2 *Clustering*

Clustering is an application of AI techniques in which the output decision performs the task of dividing all of the input data into one or more clusters, where each cluster contains those inputs that are close according to some definition of closeness. If we contrast the process of clustering to classification, during the learning stage of the life-cycle, the clustering model does not require the definition of pre-existing classes. During the inference phase, clustering takes an input and identifies the cluster which is closest to the input.

There are many different types of clustering algorithms, some look at the set of neighbors that are closest to a point, others look at measures of density of points to determine how to define different clusters, while yet others may try to identify relations and linkages among different input points.

There is a relationship between clustering and classification, in that each of the clusters which are identified as a result of the training can be considered as definitions of different classes. Thus, clustering can be used as a tool to assist people in preparing and defining labels for classification. This is in addition to using it to map or classify any input into the nearest cluster.

Clustering algorithms can be based on how close the neighbors of an input point are, how much the density of the points distributed closely together is, how well the input points match to some assumed statistical distributions or how well connected the different input points are, based on a definition of connectivity.

One specific type of clustering is association mining, in which input objects that share common attributes are clustered together. If the relationship among different input objects can be represented using a graphical relationship, then graph clustering algorithms [20] can be used for association mining. In a graphical relationship, information is represented between nodes that denote different entities and links connect nodes together. Graph clustering algorithms find the nodes in the graphs that are similar to each other.

Clustering techniques can be used for many types of business operations, e.g. to understand images to distinguish among different types of tissues in medical images, to analyze network traffic to understand different modes of behavior of different devices, to model topics which may be prominent in the transcripts of customer support calls, identify devices that may be operating in a similar manner in networks, etc.

1.8.3   *Anomaly Detection*

In many types of business operations, there is a concept of normal operation and when something is not normal, i.e. an abnormal or anomalous condition may be taking place. Anomalous behavior can be used to identify potential security attacks that may be made on the network of an enterprise, a machine that may have some problematic situation and may need maintenance, a customer that may be very unsatisfied, an attempt at a fraud, and several other use-cases.

Anomalous behavior can be viewed as being related to both clustering and classification. It can be viewed as an attempt to classify all input into two classes, normal and abnormal. It can equivalently be considered as identifying input points that do not fall close to any of the clusters identified during clustering. However, anomaly detection can also be done without using any clustering or classification approaches, such as examining a system during operation, determining a baseline of measured values and then reporting when the values start to fall outside the range observed during base-lining. Anomaly detection can also be done by comparing an input to a set of cohorts, where cohorts are chosen according to some criteria, and anomalous behavior is flagged if the input does not conform to the norms defined by the behavior of the cohorts.

Anomaly detection can be used as part of preparing data for other types of machine learning, in that outliers or anomalous data can be removed from the training set.

1.8.4 *Mapping*

A mapping is a transformation of input into output where the inputs and outputs belong to two different domains, where the two domains are two distinct ways of representing the same entity. As an example, a word phrase may be expressed as a sound signal or in a textual representation. A translation of sound samples to a textual representation would be a mapping. Translation of information across different representations, e.g. translation among different languages where the same text is represented into different languages, is an example of mapping.

In many cases, mappings are done in order to improve the speed of operations. Finding similarities between strings of texts requires complex comparisons between sequences of letters that can be slow and inefficient. However, if words are mapped to a representation of mathematical vectors in a vector space, the comparison of similarities becomes more efficient. As a result, many text processing applications relying on mapping words to vectors, sentences to vectors, paragraphs to vectors, etc. Similarly, sequences of objects can be represented as vectors, and vector concatenation can be used to provide automatic composition of business work-flows.

Mappings can also be used in systems and network management applications, e.g. events observed in the network traffic or system logs can be mapped to underlying root cause which may be causing those events. AI-based techniques to map events to root-causes can help ease the burden of system and network management.

The mapping operation is related to classification, in that one can view each of the output values in the mapped representation as a class to which the input is being mapped to. The algorithms usually associated with classification would work well when the number of classes are small. However, as the number of classes become very large, mapping algorithms may be more efficient and perform better.

Mappings are used widely in many AI-enabled processes as components that help the end-to-end operation execute in a better manner.

1.8.5 *Filtering*

The term filtering, originating from the field of signal analysis, but used in many business processes, refers to the act of removing unwanted components or features from a signal. Many business processes are driven by signals monitored via sensors, and decisions are made by analyzing the outputs from those sensors. The signals from sensors are usually converted into events that are fed as input into the business process. Many of these events could be redundant and duplicate, and filtering eliminates or compresses those events to a smaller subset for efficient processing.

Filters can also be used for smoothing signals and for eliminating erroneous information. If you use GPS-based navigation systems in your car, you may notice a few occasions when the system believes that your car is not on the precise road, but on a near-by off-road point. This is due to the inherent errors in estimating position using the GPS system. Smoothing the location over several previous measurements eliminates the likelihood of such errors.

Another use of filtering is to make indirect measurements of variables that may not be directly observable from the data itself. One type of filter, the Kalman Filter, uses information about the model of the system, a series of possible noisy measurements and configuration of the system to estimate the state of a system, and is widely used in many areas including control of autonomous vehicles and drones.

1.8.6 *Function Modeling*

A very general type of AI function is the task of estimating a function that models the relationship between the input and the output. The goal is to come up with a function that can predict the output given the input.

In many cases, the input consists of several components, e.g. the different parameters that make up the attributes of credit card or loan applications, and the output is the decision made in the past, e.g. whether the loan was allowed or disallowed, and the task is to estimate a function that can predict this decision on the basis of the input values. Once this function is estimated from the historical data, it can be used to make decisions in the future.

In some cases, the input is time-dependent, i.e. it consists of a sequence of measurements over time. In those cases, function estimation can include predicting the future value of a measurement based on the past values and the predicted values can be used for decisions. As an example, the amount of requests seen at a website can be used to predict the estimated request rate in the future, and the estimate used to allocate the resources, e.g. the number of servers required to handle the estimated workload.

Function modeling can be viewed as the most general function of AI and most of the other AI functions can be redefined as specific cases of function modeling.

1.8.7 Goal Attainment

Goal attainment is a specific function in AI in which a goal for the system is defined, and the system is expected to obtain that goal automatically with minimal human input. Both reinforcement learning and Automaton learning using finite state machines are designed explicitly to attain a goal, but other AI models can also be used to attain a goal that is defined.

One specific example of a goal attainment would be for the system to win at a game, whether the game is Chess, Go, or Checkers. In order to attain that goal, the system can be trained to model the impact of different moves made in a goal by itself, its opponent (or opponents), and to select the best possible move that will allow it to eventually win the game.

Another example of a goal attainment would be the problem of system optimization, where the goal provided to the system is to maintain some property of the system while maximizing or minimizing some attribute of the system. An example would be to ask a data center controller to minimize its power consumption (e.g. by using the least possible number of servers) while ensuring that all applications are running with a desired upper threshold on their response time.

1.9 Summary

In this chapter, we have provided a very high level overview of AI and its usage within the business processes. We have modeled the business operation and business decision making and introduced the learn-infer-act cycle as the basic mechanism for use of AI in business. We have provided a broad overview of the type of AI models used in practice, and a taxonomy for how these models can be trained, and an enumeration of some of the generic AI functions that can be used.

In the next chapter, we will look at the challenges associated with AI model building which drive the need for federated learning.

Chapter 2

Scenarios for Federated AI

In Chapter 1, we covered the broad outlines of an AI-based business operation and discussed how it can be implemented using the Learn→Infer→Act cycle. In this chapter, we expand upon the spatial aspect of how the Learning and inference process happens, which was briefly discussed in Section 1.7.

Most real-world situations in businesses deal with systems that are distributed over a wide area and interconnected by a computer communications network. The different components may be located in different parts of the world connected by the Internet, or be within different sites of a company connected by the corporate intranet. We would group these sites into one of three categories (a) edge sites (b) proxy sites and (c) central site.

The exact definition of edge, proxy and central site would depend on the specifics of the business. For example, if we consider a manufacturer with several production plants. Each of the plant locations would be the edge site. The manufacturer probably has a data center which is the central site. There is no proxy site in this configuration, which consists of several edge sites and a central site.

If you consider an international bank with many branches in different countries, each branch of the bank is the edge site. The branches may be sending their data into a country-specific data center which will maintain the data in those centers according to the required data regulations for that site. These branches can be considered as the intermediary sites for the bank. The bank may also have an international data center where they do the bulk of their software development and maintenance. That international data center is the central site.

Another business, e.g. an insurance company, may have decided to host all of its operations at a cloud service provider. The branches of the insurance company generate various transactions that the agents of the company create for the insurance processing. The branches are the edge site whereas the cloud hosted location of the insurance company becomes the central site.

AI is data driven, and each of these sites could have five possible roles to play for data processing. A site could have (a) data generation role (b) data collection role (c) model training role (d) inference role and (e) action role.

A site with the data generation role generates the data that is required for the business process. A site with data collection role maintains the historical data generated within the enterprise. Sites with model training role use machine learning algorithms to convert the data to AI models, and need to have adequate computational capacity along with any hardware accelerators required to speed up the task of model creation. Sites with Inference role use the data that is generated to reach a decision and sites with action role are the locations where the recommended action is taken.

During the Learn phase of the Learn→Infer→Act cycle, sites with data collection role and model training role are the ones that are active. The site with data generation role may be active in some cases, if it is generating the training data, and in other cases it may not be active, e.g. if the training data is being acquired from some other source. Sites in the the data collection role collect an adequate amount of training data, either in real-time or previously collected data. If the site does not have the role of model training, the collected data is sent over to the site with the model training role. That site would process the data to create an AI model.

The model that is created at site with the model training role is sent over to the site with the inference role, which is activated during the inference phase of the Learn→Infer→Act cycle. During the inference phase, the other sites that are active are the ones with data generation role. Data created at the sites with data generation site is sent to the site with inference role which will convert the generated data into a decision. The site with data collection role may or may not be active in this phase depending on the use case. If the use case is collecting data during the inference phase to update the model later, the role of the data collection should be active. However, if the use case is such that inference data can not be collected, sites with this role may not be active.

Once the decision is made, it is sent over to the site with action role, which would take the action. Of course, if the inference site and the action site are the same, any transfer activity can be avoided. The different roles of sites and when these roles are activated during different phases of the cycle are shown in Figure 2.1.

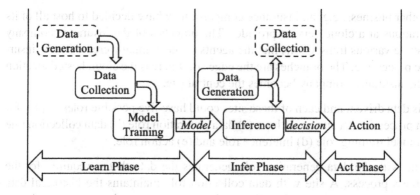

Figure 2.1: Active roles during different phases of the Learn→Infer→Act cycle.

The assignment of the different roles to edge sites, proxy sites and central sites leads to different patterns for enterprise AI. In some of these patterns of enterprise AI, we would need to use federated AI techniques.

2.1 Abstracted Patterns for Enterprise AI

Each pattern of the enterprise machine learning provides a different mapping from the five roles of data generation, data collection, model training, inference and action to the edge, proxy and central sites. In almost all cases, the role of data generation and action would fall upon the edge sites. The role of data collection, model training and inference could be assigned to any of the other three sites.

2.1.1 *Centralized AI*

The conventional approach for machine learning in the industry is that of centralized machine learning. In centralized learning, the roles of data collection, model training and inference are all assigned to the central site. As in all patterns, data generation and action roles are assigned to the edge sites. A proxy site need not be present in the centralized learning, or the proxy site may be present with the additional role of being a data collector only during the learn and infer phases.

The use of the central site to train the model has several advantages. It would allow the use of a system which has customized accelerators to speed up the training process. It simplifies the task of updates to the model, if needed in the future. The central site is likely to have the computational capacity needed for model training, as well as the human expertise to optimize and make the best machine learning model. The training data that is collected can be curated and cleaned up at that site.

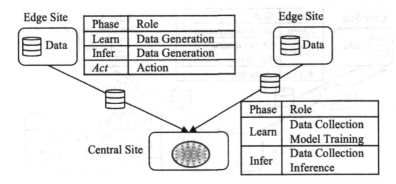

Figure 2.2: Centralized AI pattern.

The abstract pattern of centralized AI without any proxies is shown in Figure 2.2. The data for creating the machine learning model is being curated from many edge sites that have the role of data generation. Two such edge sites are shown within the figure. The data that is generated from each of the sites is sent over to the central site. The central site is also the data collection site.

During the learn phase, the edge sites send their generated data over to the centralized location to create the AI model. The model is created, maintained and updated at the central site. During the infer phase, data generated at edge sites is sent over to the central site where the pre-trained model is used to determine the right decision to undertake. The resulting decision is sent back to the edge site where the action is taken. The role of data collection during the infer phase will be done by the central site.

In general, some of the edge sites may take on the roles of data generator only during the learn phase, some of the edge sites may take on the role of data generator only during the infer phase, and some of the edge sites may take on the role during both the phases. However, for Figure 2.2, we assume that all the edge sites have this role in both the phases.

In this pattern, there is movement of data from the edge sites to the central site. However, the model never moves out of the central site.

2.1.2 *Edge Inference*

In the edge inference pattern shown in Figure 2.3, the central site has the role of data collection and model training during the learn phase of the Learn→Infer→Act cycle. However, the model that is trained is moved back to the edge sites, who have the role of inference during the infer phase of the cycle. The proxy sites may not be present, or if present, have the role of data collection during the learn and infer phases.

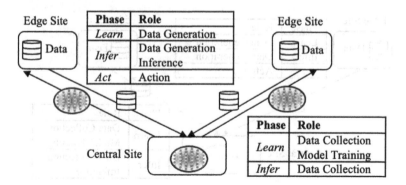

Figure 2.3: Edge inference pattern.

The primary difference from the centralized learning pattern is that the model moves out from the central site to the edge site between the learn phase and the infer phase. This has the advantage that, during the inference, the network latency between the edge site and the central site can be avoided. This allows for a more responsive decision and action during the inference stage. If the network between the edge site and central site is expensive, this approach can save significantly on the cost of the network during the infer phase.

In some use-cases, the connection between edge sites and the central sites may be intermittent, and can not be relied upon during the infer phase. In those cases, the edge inference pattern is a viable approach while the centralized AI pattern would not work. An example would be airplanes or unmanned aerial vehicles that may need to operate without a high bandwidth network connection available to them during their flight. They can load the model from a central site before their flight, and operate independently performing inference using their local models during their flight.

If the model needs to be updated, the data needs to be collected and sent back to the central site. The role of data collection during the infer phase can be taken on either the edge site or the central site. The edge site may do the data collection at a slower time-scale than the decision making required during the inference stage. As an example, the edge site may collect the data that is generated and upload it to the central site at a later time, e.g. during off-peak hours when the network may be less congested or less expensive.

2.1.3 *Federated Edge Inference*

The federated edge inference pattern is a modification to the edge inference pattern, with the difference that some or all of the edge sites exchange information with each other during the infer phase of the Learn→Infer→Act cycle. They may

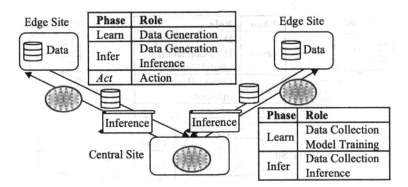

Figure 2.4: Federated edge inference pattern.

choose to do this with the assistance of the central site, assuming they have connectivity with the central site, or operate without any assistance from the central site.

Figure 2.4 shows this pattern when the central site is available during the inference cycle. Part of the inference role is assigned to the central site which is assisting the edge sites in the inference role. An example of federated edge inference can be seen in checking for cracks in a bridge using drones. A swarm of drones may be dispatched to examine the condition of a bridge. They may be searching for cracks on the surfaces, and if a drone finds an area which is suspected to have some cracks, it may ask other drones to examine the surface as well. If different drones are carrying different types of sensing equipment, checking for the inference results from different drones using different sensing equipment can lead to a better overall inference.

The drones may operate with or without the involvement of the central site to coordinate their inference. The use of a central site makes the task of coordination simpler, but also requires a good network connection between the drones (edge sites) and the central site.

2.1.4 *Edge Learning*

The edge learning pattern is useful when the network connectivity between the edge sites and the central site is not very stable, expensive or not fast enough. In this pattern, the edge sites do not send any data to the central site during any phase of the Learn→Infer→Act cycle. The edge sites take on the role of data collection and model training during the learn phase, the role of data collection and inference during the infer phase, in addition to the usual role of data generation during the learn and infer phase, and for action during the action phase.

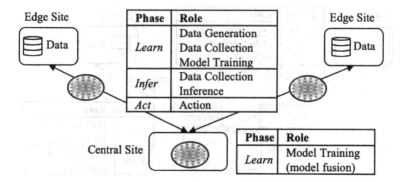

Figure 2.5: Edge learning pattern.

This pattern is shown in Figure 2.5. In this pattern, the edge sites do not send raw data to the central site. Instead, the data is converted to an AI model at the edge site itself. The AI models are moved to the central site, where the models from different edge sites are combined together into a single model obtained by fusing all of the models together. In this pattern, the role of the central site during the learn phase is to help the fusion of the models produced by different edge sites. The central site has no role during the other phases.

Edge Learning would require sufficient computation power at each of the edge sites. Moreover, the algorithms for fusing the models together would tend to be more complex than of simply training the model on available data.

2.1.5 Proxy Learning

The proxy learning pattern is useful when the edge site does not have enough computational capacity to train models, or may not have sufficient data to train a good AI model. The quality of an AI model is critically dependent on having sufficient training data. If at the same time, the edge site can not send data over to the central site for any reason, it would be useful to have a proxy site which can perform model training for the edge sites.

This pattern is shown in Figure 2.6. The roles assigned to different sites are also shown in the figure. There are two proxy sites shown, although there could be several other proxy sites. The proxy sites would be training the model from the data that they collect, and exchange models only with the central site. The central site would be combining these models together. The edge sites are effectively reduced to the role that they had in the centralized learning patterns, while significant parts of the role that they would have taken in an edge learning pattern are being done by the proxy sites.

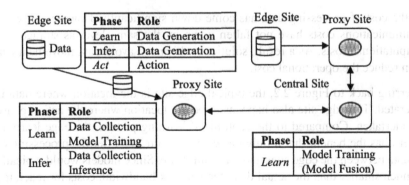

Figure 2.6: Proxy learning pattern.

At the beginning of this chapter, we described the scenario of an international bank. Each branch of the bank was an edge site, in-country data centers were the proxy sites, and an international data center was the central site. If regulations restrict the transfer of banking data outside the country, the proxy sites can train local country-specific AI models. The central site can combine those models to create a global model incorporating the patterns present in data across all countries.

The edge learning patterns and proxy learning patterns show an approach for model building which requires federation during the learn phase. The federated edge inference shows an approach which requires federation during the inference phase. We can refer to these three patterns as typical approaches for federated AI.

In the next section, we look at some of the motivations why federated learning and federated inference may be needed.

2.2   Motivations for Federated Learning

There are many factors that drive the need for federated learning. These factors include the cost of operations, network constraints, regulatory restrictions, along with data privacy and security concerns.

2.2.1   *Operational Cost*

In most enterprises, a tremendous amount of data is generated during the normal course of operations. In order to analyze and process this data, they would need to move the data to some location which has sufficient computational and storage horse-power to process that data. Both the movement of data, the storage of data, and the computation power required to process it add to the cost of managing and operating it. With advances in storage technologies and computation capacity, the cost of processing data has come down significantly. However, network communications costs have not fallen down by the same level as storage and computational costs. As a result, solutions which can avoid data movement can often reduce the operational cost.

Referring back to Figure 2.2, the typical edge site is the location where data is generated. The edge site also happens to be the location where the action phase is undertaken. Compared to the centralized learning pattern, the edge learning pattern has the benefit that it only moves models created after the processing of the data between the edge sites and the central site. Since models would typically be much smaller than the actual data, the network bandwidth costs for federated learning can be significantly lower. The same benefit applies for the proxy learning pattern except that the reduction is in network bandwidth between the proxy sites and the central site.

Moving a smaller amount of data to also has the side-benefit of reducing the amount of processing power required at central site. In general, model training is much more computationally expensive compared to fusion of models. This reduction in cost at a central site will be offset with an increase in computational needs at the edge site or proxy site to train the model.

In many situations, the computation capacity needed is already available at the edge sites. In these cases, edge learning (or proxy learning) provides an obvious cost advantage. The cost advantage can be more pronounced when the central site has a cost associated with additional processing power, e.g. in cloud computing, costs are paid according to the amount of resources used. If an enterprise is using cloud resources and has access to computing at edge/proxy sites, edge learning or proxy learning can save significantly on the costs.

However, if new capacity needs to be placed at an edge or proxy site, the costs of providing that infrastructure needs to be balanced with the savings in the operational cost at the central site. In those cases, edge learning/proxy learning may not necessarily save on the operational cost. The final answer would depend on the relative costs of compute, connectivity and storage. However, in many cases, a federated learning approach turns out to be the cheaper approach for creating and training AI models.

2.2.2 Network Constraints

In many cases, an enterprise may want to use federated AI because the network connectivity between the edge site and central site is constrained. The constrain may manifest itself as limited bandwidth, a high latency or low reliability. With constrained network connectivity, it may not be possible to move data in the raw from the data generation site to other sites. The better approach, and in some cases the only feasible approach, is to process the data locally at the edge site.

This situation is very common when the edge sites are in remote locations, e.g. in the forestry industry or in the mining or off-shore rigs. In the forestry industry, the edge site would be a ruggedized computer installed in logging vehicles that are operating in remote locations. Off-shore rigs often have connectivity only via satellite networks. Satellite links have relatively high latency, which could be around 500 ms round-trip time as compared to 150 ms for cellular 4G networks or 40 ms for wired networks depending on the location. The bandwidth available on satellite links is also much smaller, typically about 1 Mbps per second would be available in contrast to wired infrastructure which can easily provide speeds of 1 Gbps at a very low cost.

Another environment with heavily constrained networks is in the travel and transportation industry. The ships on the sea can have computing capacity on-board, but they have very limited connectivity to the rest of the world. Airplanes, when they are in flight, also have very constrained connectivity with the ground infrastructure, although the connectivity is much better when they are at the airports.

In military operations, network connectivity is invariably constrained. This is because most military operations happen in areas that are remote, have limited infrastructure for network communications, and the available communications infrastructure is always at the threat of being disrupted due to enemy actions. Modern military also consists of ships, planes, helicopters and drones, all of which would have very limited network connectivity when they are in operations.

Network constraints can be a problem even with wired networks. Cellular networks, as an example, have a tremendous infrastructure for moving data around. However, due to the large geography at which cellular network operators need to serve, cellular network providers in large countries operate a dozen or more data centers to maintain call and connection records of their customers. Cellular network operators in United States would have a dozen or more data centers. In India, cellular networks are divided into twenty-two circles, and it is typical for a cellular network carrier to have a dedicated data center for each circle. The volume of records that is generated every day at each of these data centers are huge, which makes collecting all the data at a single site time-consuming, despite the existence of good wired network links.

When data movement is a problem, patterns supporting federated AI can provide a significant advantage over the centralized learning pattern.

2.2.3   *Data Privacy Regulations*

In some cases, regulations may prevent the movement of data from the edge site to the central site. This is very often the case in industries with a high degree of regulation, such as health-care or finance.

As an example, a hospital chain may operate several clinics where patients visit their doctors, and may also have research centers where research scientists try to analyze patterns for spread of diseases or study the effectiveness of different types of cures. Patient privacy laws in many countries may prevent the sharing of raw patient data to the research scientists. While health-care professionals at clinics may be able to maintain their patient data and view it without restrictions, the research scientists may not be allowed to view this data. If the research scientists want to build a model that is based on the data present at different clinical sites, federated learning might be able to provide them with a viable approach.

Similar restrictions on the sharing of information arises in the case of government agencies. Different agencies are governed by different regulations, and they may not be allowed to share data freely with each other. As an example, tax records in the United States can only be shared with a selected set of other government agencies. Similarly, some states in the United States may prevent access to some information, such as driving records, to some federal agencies. These regulations must be complied with, yet they may become a hindrance when the different agencies need to cooperate for an activity of mutual interest.

2.2.4 Security and Trust Considerations

In some cases, security considerations may prevent the movement of data from the edge site and central site. This situation frequently arises when the different sites are owned by different organizations.

In the current computing environment, the use of cloud computing technology [21] is very popular as a cost-effective and elastic approach for obtaining a variety of data analysis services. However, this would require handing control over to an infrastructure managed by another organization, namely the cloud service provider. In that case, some of the data which is considered sensitive many not be sent over to the cloud site, and needs to be processed locally.

Another common situation where different sites may be owned by different organizations occurs in the consortium settings, where many organizations are planning to work together and to collaborate on a common task. However, sharing of data among all of the consortium members may not be completely open, and organizations may be worried about the security of some parts of their data. In these cases, members are willing to share some of their data with others, but not share all of the data.

In general, different sites are willing to share information with each other because they have some level of trust in each other, and they believe that the information sharing will be mutually helpful. However, the trust may not be absolute, and in those cases, transforming the data so that they will be not completely visible to the other party, e.g. by transforming them to a model, would provide a better approach. Federated learning allows parties with limited trust in each other to share insights gained from an analysis of their data.

2.3 Consumer and Enterprise Federated Learning

While federated learning can be useful in many different contexts, it makes sense to differentiate among two broad categories of federated learning, each of which need to address very different set of problems and challenges. We define these two categories as consumer federated learning and enterprise federated learning.

2.3.1 *Consumer Federated Learning*

Consumer federated learning refers to the situation when the data generation happens at a device which is owned by a individual as opposed to an organization. The most common instance of such a device is a smart-phone. A smart-phone contains a significant amount of data which can be sensitive, such as the location and location history of the user, their private information such as birthdays and family members, their web browsing history, their social networking history, etc. This information can be very useful to many types of mobile application developers, who can mine the information to improve their services. These improved services could improve the accuracy of predicting what users will likely type ahead in their web-searches, what types of movies they would like to see based on their history, or what types of product recommendations they ought to be getting.

The typical configuration for consumer federated learning is shown in Figure 2.7. The configuration consists of the site where AI model fusion happens, and several smart-phones which are essentially the edge sites. Mobile applications on the smart-phones would generate the data from their local consumption. All the smart-phones would typically be running the same application which should be

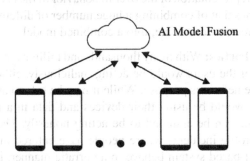

AI Model Fusion

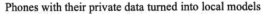

Phones with their private data turned into local models

Figure 2.7: Consumer federated learning.

generating data in a consistent and common format designed and implemented by the developer of the application. For any popular smart-phone application, the number of smart-phones running that application could easily run into thousands or millions.

In the current environment, it is common practice for many search engines, social networking sites, and retailers to collect this data at a central site and mine them for common patterns across all of the users. However, there is also a growing sense of angst at the violation of the privacy that occurs when sensitive information is collected and analyzed by a service provider. As a result, there has been a push to create technologies that will allow the exploration of common patterns across different users without requiring then to move their data to a central location. Federated learning techniques provide an approach to perform this task.

With the federated learning approach, the consumer devices would not need to move their data to the central location. Instead the models would be built on their local devices, and these models combined at the back-end data sites of the service provider. Consumer federated learning is characterized by the presence of a large number of data generation sites (or rather devices), a common format for data being analyzed defined by the developer of a mobile application which embeds federated learning techniques, and operation on a smart-phone.

However, in order to build this type of service, the federated learning approach needs to address a few fundamental problems. These problems include:

- **Small Data Size**: The service provider would typically be providing its services to millions of consumers, or at the very least to a few thousands. In these cases, the overall data that is to be mined for patterns is split into units of many small sizes. In general, AI model learning algorithms are not very good at extracting patterns from data which is small in size. As a result, the model built into each of the site is highly likely to be very inaccurate representation of the overall behavior of the users. The problem then becomes that of combining a huge number of different models, each of which are very inaccurate, into a combined model.

- **Malicious Parties:** With many thousands and millions of users, it is likely that some of the users would be acting maliciously, either as a prank or with a more nefarious purpose. While it may be safe to assume that most individuals would be using their devices and data in a normal manner, not everyone can be assumed to be acting honestly. There have been a few high profile incidents where people have entered malicious data to make an AI-based system behave in an erratic manner. For example, an AI-enabled chat-bot was trained to behave in an obnoxious manner by feeding it bad data in a crowd-sourcing experiment [22], computer vision algorithms can be fooled by manipulation of the input images [23], and similar attacks can be launched on speech analysis systems [24] . While these attacks are not directed exclusively to federated learning settings, a federated learning environment is more vulnerable to such attacks when a set of users try to manipulate the training process of their individual machines.

■ **Limited Computation Power**: Because training of models is a compu- tationally intensive process, the training of models on phones may have undesirable side-effects such as draining the battery, or consuming too much of the available computation capacity. While data availability is limited, the requirement to not interfere with the normal operations of the phone may also restrict the choices of AI models that can be practically done on the smart-phone.

Because of these challenges, the federated learning on the consumer devices have not seen a significant uptake in its adoption in any of the widely deployed consumer applications. A significant number of researchers have been working to address the problems associated with the consumer grade federated learning [25], so they might still find some adoption within real-world use-cases. However, the enterprise federated learning, which we are defining next, is more likely to be the area which sees the first practical applications of federated learning technology.

Enterprise federated learning refers to the situation where the data generation sites are owned by an enterprise. These sites generally happen to be an office, a manufacturing plant, or other environments where data is generated by many different devices, and is collected and stored into some type of database at the data generation site. The typical set-up of enterprise federated learning in shown in Figure 2.8. It shows several sites, each with a large amount of data stored in databases, which all need to be mined in order to create the overall AI model. Each site could be an individual building of an enterprise, or it could be a data center, or other logical partitions of data maintained within an enterprise, such as data lakes, or data warehouses. These various terms refer to portions of the enterprise

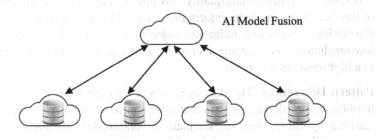

Figure 2.8: Enterprise federated learning.

IT infrastructure which is responsible primarily for storing and maintaining data that is generated in normal course of business operations.

In contrast to consumer federated learning, which consists of millions of devices, the number of sites in enterprise federated learning would be much smaller, typically ranging in single digits to a few tens of locations. Each of these sites would typically have a substantial amount of data to train and learn their models. Computational capacity is also not an issue because the site would typically have servers with adequate capacity to train the models. Note that the fusion site itself may be co-located with one of the sites containing local data. This co-location would be the typical configuration if the AI models are all trained within the data centers owned by an enterprise. In some cases, the fusion site may be located in a different physical site and not may any local data. That would be the typical configuration if a cloud-hosted service is utilized for creating the federated AI models.

2.3.2 *Enterprise Federated Learning*

Enterprise federated learning has its own set of unique challenges. These include:

- **Heterogeneous Data Formats**: Because the enterprise data sources may have been collected independently, the data may have been collected in many different formats, even if they all refer to the same type of data. Suppose the data refers to Human Resources personnel records, and it is maintained by a company which acquired another company. The records may be maintained in different formats, even though they may refer to the same type of data.

- **Varying Data Quality**: Different sites may have different practices for recording and storing data. As a result, data maintained at one site may be well curated and have a very high quality while data maintained at another site may have a lot of errors or be missing portions of the record. Sometimes, the difference in data quality may just be because of the differences in the format in which the information is stored, e.g. a location may be storing information like audio and video in a format that supports only low-resolution content, while another location may be storing information in a high-resolution content.

- **Pattern Differences**: The data at each site encodes some patterns, i.e. implicit relationship in the data that can be minded by an AI or machine learning algorithm. However, the patterns contained in different sites can be very different. As an example, a site containing banking transactions for the East coast of the United States may show some patterns for banking activities that may be different those in the middle of the country. When combining data from both sites, care must be taken so that one site's patterns are not combined improperly. If the east coast data primarily refers to transactions done in an urban environment whereas the middle of the country has transactions in the rural environment, combining the transactions may lead to misleading patterns.

■ **Trust Differences**: In many enterprise environments, the trust among different organizations may not be complete. This may result in restrictions on how data is moved across the sites. The difference in trust may be caused due to regulations prevalent in the industry, e.g. patient privacy regulations may limit the type of data that can be shared between the out-patient departments such as hospitals of a health-care organization, and its research department which is trying to explore studies on impact of medicine on patients. Similar trust restrictions may be in place among different agencies of a government entity. In other contexts, an enterprise may be a loose consortium of different organizations, and trust relationships among these organizations may not be absolute.

The differences in operating assumptions among these two different types of federated learning are summarized in Table 2.1. The number of sites in consumer federated learning are in a different order of magnitude to that of enterprise federated learning. As a result, each site in the consumer federated learning has a very small fraction of data and a relatively modest volume, while each site in the enterprise federated learning would have a much larger amount of data. The typical device for consumer federated learning would be smart-phones whose computation capability and storage capability is significantly less than the typical device for enterprise federated learning, which would be a computer server. Malicious devices can not be ruled out in case of consumers, but business arrangements would establish limited trust among the sites in enterprise federated learning, which would also prevent malicious behavior. The data format would be the same in consumer federated learning, but is likely to be different in enterprise federated learning.

As can be seen from Table 2.1, the challenges of enterprise federated learning and consumer federated learning are very distinct. The focus of this book is

Table 2.1: Comparison of consumer and enterprise federated learning.

Aspect	Consumer Federated Learning	Enterprise Federated Learning
Number of Sites	thousands-millions	under hundred
Typical Device	Mobile Phone	High-End Servers
Trust Relationship	Untrusted	Limited Trust
Malicious Behavior	Expected	Unexpected
Data Volume at each Site	Tiny	Large
Data Format	Consistent	Inconsistent

on enterprise federated learning, which arises in many different scenarios, as described in the next several sections.

2.4 Enterprise Federated Learning Scenarios

There are many situations under which an enterprise may want to deploy the power of federated learning and federated inference. These situations will be driven by the motivations described in Section 2.2.

Whenever there is a distributed business with restrictions on movement of data, a scenario for federated learning is created. In Section 2.2, we alluded to some of those scenarios, which include but are not limited to:

- Forestry and Logging: where many distributed assets are located in remote areas with hardened computers on their outdoor equipment, but with limited and expensive network connectivity.

- Multi-national Companies: which have data centers in many different countries, but have restrictions on moving data across the national boundaries.

- Manufacturing Plants: which may have many manufacturing facilities, and moving data from the manufacturing facility to a central site is expensive.

- Retail Stores: which many have shopping outlets with computing equipment present in the back-office of the retail store, and want to continue without being critically dependent on network connectivity.

Additional scenarios where federated AI can be used successfully involve situations which have intelligent devices moving around and the mobility makes communication networks weak in some locations. These include intelligent cars with on-board computers, drones or unmanned aerial vehicles (UAVs), ships and planes, trucks moving on the highways, etc. These scenarios will arise in both the military domain, where military planes, ships and unmanned assets are mobile, as well as civilian (non-military domains). Each of these scenarios have sufficient computational resources at the edge site to train a model, and some motivation for not moving that data to a central location.

In the following subsections, we describe some additional scenarios.

2.4.1 *Subsidiaries and Franchises*

Many enterprises are arranged as loose conglomerations of subsidiary companies. It is a fairly common arrangement for multi-national companies. Generally, such companies may be organized as one company per country, subject to the rules of the country they are operating in. The complete ownership, or in some cases part of the ownership of the subsidiary in any country other than the home country of the enterprise is held by the company in the home country. While the subsidiary may be completely owned by the parent company, it still may not be able to retrieve data from the subsidiary because of prevailing regulations.

This scenario is very common for companies that may have the parent country in the United States and may have subsidiaries in European Union. In general, Europe has more strict regulations about how personal data is maintained and handled than the United States. As a result, movement of raw data about company clients in Europe to the United States may be restricted. In those cases, if the parent company would like to build a model characterizing the patterns of its clients' behavior in Europe, it would not be able to build the model by moving the data across to the data centers in the United States. However, if a model is trained locally, and only the model is moved to the United States, it may be able to satisfy the regulatory requirements of the local country.

The typical setup of a subsidiary is as shown in Figure 2.9. The larger conglomerate would own interests in multiple subsidiaries. Each subsidiary would be responsible for its own operations, and would generate its own data. Each subsidiary needs to run its own IT operations. It is common for the conglomerate to also have a central IT operation. Some of the IT functions may be controlled centrally and performed by the conglomerate, while some of the IT functions may be controlled independently by the subsidiaries.

In some environments, the data format can be assumed to be the same for all of the different sites. They would be defined and determined by the central IT operations of the conglomerate. However, the volume of data that may be collected for each of the different sites may be very different. Not all subsidiaries would have the same volume of transactions. This difference in the volume of data needs to be accounted for when performing the task of federated learning. One would like to extract patterns that are more clear and easier to extract when data volumes are

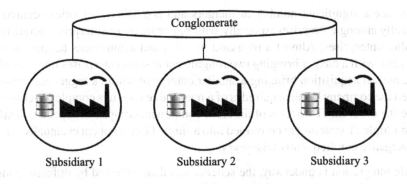

Figure 2.9: Subsidiary and franchise setup.

large. At the same time, if there are unique patterns in the data of a subsidiary with a smaller volume of data, it would be useful to extract those patterns, and not have them drown in the volume of the subsidiary with the larger amount of data.

The characteristics of data available at different subsidiary sites may vary significantly, e.g. one subsidiary may have mostly young professionals as clients, while the other subsidiary may have mostly older retirees as clients. When the subsidiary sites are in international locations, their clientele and operations may be influenced strongly by the demographics of the population that they may be serving. Thus, each site may be collecting data containing some patterns that would be shared and be in common with those of the other subsidiaries, while they may also have some patterns in their data that are unique to their environment and context.

A franchisee also looks very similar to a subsidiary, except that a franchise would typically be smaller and have more independence in their operations. Because of the small scale of operations, franchises may tend to rely more upon the IT services provided by the IT personnel of the conglomerate. However, some franchises may chose to have their own IT operations, specially if they want to have independent records and not rely on the conglomerate for part of their IT operations. When franchises operate with their own IT department, they may be deviating from the format and conventions used by the central IT office of the conglomerate, and functions. Franchises that are handling their own operations independently may want to protect their business data by keeping it private to themselves. This need for independence is more likely to be seen when the franchisee is operating a number of offices in an area. As an example, the conglomerate may be a fast food chain, and the franchisees may be operating several stores within a select geographical region under license and business agreement with the conglomerate.

2.4.2 Mergers and Acquisitions

There are a significant number of mergers and acquisition activities occurring annually among enterprises, specially with big or medium enterprises acquiring smaller enterprises. Almost a thousand mergers and acquisitions happen annually [26], with a merger bringing two roughly equal sizes enterprises into a single unit, and an acquisition bringing a smaller enterprise within a bigger enterprise. When the company being acquired is of a reasonable size, it is complex to consolidate the different IT systems of different component companies together. While the multiple IT systems are converged into a single IT component eventually, such convergence can frequently take years.

While integration is underway, the schema and data collected by different components of the company could be very different. In these cases, the schema of the

Figure 2.10: Mergers and acquisitions setup.

data collected at different sites may be different, in addition to the other challenges faced by an enterprise consisting of multiple subsidiaries. In some cases, the acquisitions may be required to maintain their operations to be independent and separate even after they become a single entity. As an example, a company which operates both a pharmacy business and has acquired an insurance business may need to maintain a separation between the data of the two sides of acquisitions.

If we compare the structure of the acquisitions/mergers with that of the subsidiaries, the relationship among different entities is much more of a peer relationship compared to that of the subsidiary/franchisee. Even though the acquiring company, shown as the Main Corp in Figure 2.10, may be in a stronger position to control the operations of the companies it has acquired, it may often find the cost associated with changing the operations of the acquisition to be relatively expensive. Depending on the level of integration among the main corporation and its acquisitions, the data maintained in different acquisitions may be maintained separately. Over the long term, it is possible that all the IT processes and the associated data would be available at a single central location controlled by the main corporation. However, until that eventual integration, there would be many different independent silos of data and information in the overall business.

The main corporation may choose to copy data from its acquisitions into a central data warehouse or data lake in order to analyze and explore patterns across all of their data. However, since such data movement is often expensive and time-consuming, and in some cases not viable due to regulatory restrictions, federated learning provides a mechanism for patterns to be extracted from data that is distributed in many different locations.

2.4.3   *Outsourced Operations*

In modern business, outsourcing of operations is a common practice. Many companies focus on operations that they consider their core competency for their business and outsource all other operations to other companies. The support of Information Technology infrastructure required to run the operations of the business is often the subject of such outsourcing. In these cases, the company would select another company (the provider company) which has the expertise and operational knowledge to perform the IT support to manage its infrastructure. The provider would typically be providing these services to many different client companies, and would not be reliant on a single company. In other words, a single specialist provider may be supporting many different companies. In addition to IT support and operations, such outsourcing is also fairly common for many other aspects of operations, including facility maintenance, inventory and asset management and maintenance, data analysis, shipping, payroll, etc.

The provider has access to information that originates from many different clients. In general, the data from one client can be considered to be proprietary to that client, and it would be poor business practice for the provider to share data of one client with another. However, the provider can also gain significant advantage from learning patterns that are common across the data of all of its supported clients, and use that to improve its own business process.

As an example, let us take the case of management of physical assets of a company, a task which is often better delegated to another company that can run a cloud-based asset management service. The physical asset management service would keep track of assets owned by the company, records of their service and maintenance, and when they had a failure. A typical arrangement used by the provider company is shown in Figure 2.11. The servers used to support different clients are separated and made secure by means of firewalls that can insulate the different clients from each other in the cloud center or the data center of the provider. The management system is used by the provider to support the operation of the various servers of different clients, and is itself protected by firewalls. A virtual private network may be set up between the campuses of each client and

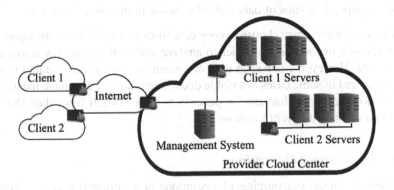

Figure 2.11: Outsourced operations structure.

the servers that they may have in the provider cloud center. In a real setup, several other tiers of firewalls may be used to provide additional security.

If the provider company on the cloud is hosting the asset maintenance service of many different companies, it would be able to combine the knowledge gained from the hosting of the one company to improve the hosting processes of the other company. This would benefit all of the companies that are being hosted. The AI models for improving the services can be created at the management system of the provider by collecting all the information available from the different clients. However, companies that are working with the provider may be wary of such use of their asset information. The provider may be hosting their competition, and they may be concerned about the competition learning about their asset history. They may not want the information about their assets mingled into a general information about all the assets. However, they may be willing to let the provider develop models for them that predicts when maintenance ought to be scheduled, or what the predictive indicators of a failure might be. Sharing of these models and combining them across all of the clients of the provider might be acceptable.

In the case of cloud hosting service , a similar situation arises not with the data owned by the client, but in terms of the operating characteristics of the machines that are used to host the service. In order to provide isolation among clients, the provider may be using approaches such as different physical machines for each client, separate virtual machines for each client, or separate containers [27] for each client. The performance metrics for each of these separation mechanisms can be used by the provider to get insights about better management of the infrastructure, and automate triggers that can flag anomalous behavior of the service. However, data sharing agreements between the clients and the provider may preclude the sharing of this data with other clients, or commingling of data from different clients.

The driver for use of federated learning in this case would be the contractual restrictions that exist between the client companies and the provider, and restrict the flow of data across organizational boundaries.

2.4.4 *Telecommunications Networks*

Sometimes, the need for federated learning may arise simply due to the scale at which data is being generated, the scale being of a nature such that movement of the data to a central location for analysis would not be feasible. This happens when data is being generated too fast to be transported across the network, or the stored data is just too large to move around.

A typical setup where this situation arises is in the context of cellular networks. Cellular networks typically contain many cell tower locations, which support a variety of mobile phones. During their operations they may contain data from

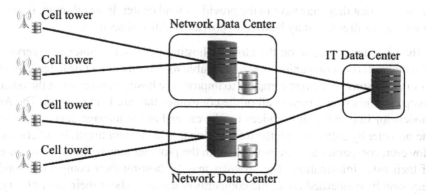

Figure 2.12: Telecommunications network structure.

many different phones, including but not limited to information about the radio signals that they are receiving from them, and variations of the signal strength and intensity. While the physical topology of different networks may be different, any data collected from the cell towers would typically be sent over to a network data center, each network data center supporting several cell tower locations. The information from network data centers may be combined together into a network operations center or be taken over to an IT data center. This results in a logical tree-like structure as far as data processing is concerned in cellular environments, which is shown in 2.12.

The volume of radio frequency signals at each cell tower is tremendous, due to which this information can not be sent over the cell tower to the network data center. The analysis or AI model that is to be done on such signals needs to be done at the cell tower itself. If a model that covers the patterns included in more than one cell tower needs to be created, then federated model building is the only viable option.

In telecommunications networks, records about the data connections or voice calls made by clients are recorded at network data centers, and copied over periodically to the IT centers for record keeping and billing purposes. A single telecommunications company may have several millions of subscribers, and may be recording billions of calls every month. They may also have different IT data centers, each IT data center covering a different geographic region. Because of the volume of the data at each data center, it is expensive and time-consuming to move the different call records together, and it would be more expedient to use federated learning techniques to extract patterns from the call records.

While these examples are given in the context of telecommunications networks, such massive amounts of data stored at many different sites can arise in many different industries for large corporations.

2.4.5 *Consortia and Coalitions*

A consortium consists of many different companies come together to share information or details in a specific field. As an example, a consortium of banks may be interested in sharing insights so as to better provide fraud detection. They may be able to share and provide models to each other, but not be able to share the entire amount of data in clear. In other cases, the consortium may be a group of medical companies, which are able to share data with each other, but only if it is anonymized and aggregated. They would like to work with each other while preserving the privacy of the members maintained within their records.

Healthcare consortia [28, 29] are a typical representation of the needs of members to maintain the anonymity and privacy of the records they maintain while still trying to share data that can be beneficial across all members of the consortium. Such a consortium may be formed among different countries in a region, and the countries may want to share information about the spread of infectious diseases in their hospitals, or share information about the efficacy of medicines to control a specific disease. The best models for the consortium may be made by combining data from all members, but sharing the data in the raw may not be feasible under the privacy regulations of member countries. However, creating models from individual data repositories may provide an acceptable alternative.

A specific instance of a consortium is a military coalition, which consists of armed forces of different countries cooperating together to attain a specific military objective. The coalition may be interested in maintaining peace in a region that may be experiencing regional unrest, or the coalition may be interested in maintaining a strategic balance of power against another coalition. Members of the coalition would share data with each other, but they may not share all data that they may have with other members. As an example, a member of a coalition may have developed technology for very high resolution imaging and may not want to reveal that fact to other coalition members. Instead of sharing raw high resolution images, the coalition member may prefer to share the low resolution images using the technology all coalition members have access to. If sharing of the images consumes too much bandwidth, the coalition members may prefer to exchange AI models since bandwidth and network connectivity for coalition members operating in remote areas is usually constrained and limited.

A typical scenario in military coalitions requiring federated learning is shown in Figure 2.13. It shows three countries, each of which have their troops located in their base camps. The task of the coalition is to monitor the area to ensure that a cease-fire agreed upon by warring parties in the area is maintained. In order to do the monitoring, each coalition member would have deployed a variety of surveillance equipment to monitor the areas around their base camps. They can automate the processing of the surveillance information by sharing data with each other so that all countries can build their AI models to analyze the collected

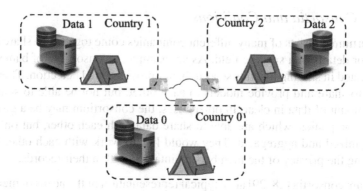

Figure 2.13: Coalition scenario.

information. However, they may not always be willing to share the data, and even if they were willing to share the data, the network connecting their base camps may not have adequate bandwidth.

Similarly, naval ships or airplanes belonging to a single country may collect a variety of data using the instruments they may have. If the ships want to analyze the collective data they may have gathered, it would be more bandwidth efficient for them to share the models, as opposed to transmitting huge volumes of data over the wireless or satellite links connecting the different ships. The same would be true for naval vessels or airplanes that are collaborating together as a coalition from different countries, with the added complexity that sensor readings may not be shared freely across countries, even if they are part of the same coalition.

A military context that is very similar to coalition operations is that of multi-domain operations. For any large country, the armed forces are specialized to operate effectively in different domains of warfare, where a domain consists of a specific area of operation. Typical domains of operations include land, sea, air, space, air and cyber. The land domain would consist of sensors, vehicles and soldiers operating on land, the sea domain would consist of ships and sailors, the air domain would consist of UAVs, planes and other air-borne vehicles, the space domain would consist of satellite signals, and the cyber domain would consist of computers and communications networks, including the Internet. In each of the domains, military of any nations would want to take on a position superior compared to its enemy forces. In multi-domain operations, the armies combine the information available in all domains to try to get an advantage across each of the domains.

Since the communication network between different domains would be constrained relative to the demands made by the data being generated in the field, extracting patterns as AI models from each domain and then combining them together provides the best approach for creating an AI model across all the domains.

2.4.6 *Regulated Industries*

Many industries operate under strict regulatory guidelines imposed by the government. The regulatory guidelines are usually put into place for a broader benefit of the community, and to prevent the misuse of information collected by the business concerns in the industry, and to ensure that the rights and benefits of the general population are preserved. However, regulatory guidelines can frequently prevent the sharing of information across different parts of an organization.

An example of an industry with strict regulations on sharing information is healthcare. The clinical records of a patient can only be shared under the restrictive conditions with the consent of the patient. While physicians may access the records to provide medical care for the patients, access is restricted to most other entities. This helps prevent the misuse of the patient information, e.g. avoid a situation where a life insurance company can access health record of its subscribers and cancels life insurance policies of any person who may have sustained an injury or accident. However, this restriction on data sharing can also preclude access for otherwise valid and beneficial purposes.

One such usage which may not be allowed in some countries is shown in Figure 2.14. A large hospital system may see patients as well as conduct medical research. Usually, the research organization would be different than the clinical (out-patient) organization. Data collected at different clinical groups may be stored separately depending on the type of access controls required by the system. An AI model in these cases would be run by the research organization, who would typically have the processing capacity and the expertise to create the models. However, the research organization would not be allowed to access the raw

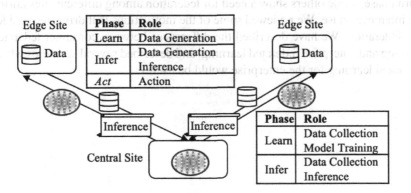

Figure 2.14: Health care company scenario.

data. The scenario shown in Figure 2.14 has two clinical sites and one analysis center, although a real hospital chain is likely to have many more clinical sites as well as several analysis centers.

One way to address these requirements would be to transform the data to hide any personal information before creating an AI model from it. However, implementing a variation of Fusion AI, which would be able to process raw data at each of the clinical site, is more likely to provide a more accurate model.

Regulations control the operation of many other industries as well. In the case of finance, there may be strict controls on sharing information about financial transactions of the clients with other entities. Among other restrictions, this might prevent sharing the records across national boundaries. Since most financial companies are multi-national organizations, this may create a challenge if they want to create a model which covers data from more than one country. Federated Learning provides one possible solution to help them create such models without transferring raw data across regulatory boundaries. Regulations also may restrict different agencies of a government from sharing information with each other. In these situations, federated learning may provide an approach for different agencies to create a model capturing patterns across all of their data, without exchanging raw data.

In general, whenever regulations prevent free exchange of data among different locations, but there may be value in building AI models that cover patterns from data across many locations, federated learning may provide a viable solution.

2.5 Summary

In this chapter, we have explored some of the patterns for deploying AI-based solutions and how different sites can take on different roles in the various patterns. Some of those patterns show a need for federation among different sites during the learn phase, while others show a need for federation among different sites during the inference phase. We reviewed some of the motivations that drive the need for this federation. We have described the differences between consumer federated learning and enterprise federated learning, and described several scenarios where federated learning for the enterprise would be useful.

Chapter 3

Naive Federated Learning Approaches

As mentioned in the previous chapters, the process of creating an AI model captures the patterns that are present in the training data. During federated learning, the patterns captured from many different locations are combined. In this chapter, we look at the basic approaches that can be used to combine the patterns from the different locations.

A very popular approach for combining models together is the approach of federation by averaging. This approach has been a significant focus of academic research [30, 31].

In this chapter, we look at the principle of federated averaging and discuss the underlying approaches behind these approaches. We refer to these approaches as Naive federated learning approaches, since they make some assumptions which may be hard to satisfy in enterprise contexts.

The system configuration under which we present the various algorithms described in this chapter and subsequent chapters is illustrated in Figure 3.1. We assume that the data for the enterprise is located in many different sites, which we refer to as Fusion client locations. In the figure, we show two fusion clients but there will be many more depending on the specific setup of the system. There is a Fusion server location which is shown as site 0. The naming of the sites as client and servers is done since we are assuming that the virtually ubiquitous client-server approach [32] for distributed computing is being used in the environment.

The configuration shown in Figure 3.1 is what would be required during the training of the AI models during the learn phase of the Learn→Infer→Act cycle

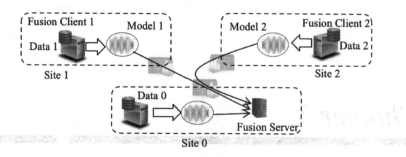

Figure 3.1: Federated learning configuration.

in the patterns of edge learning (described in Section 2.1.4 of Chapter 2) and proxy learning (described in Section 2.1.5 of Chapter 2) and during the inference phase of federated edge inference (described in Sectiion 2.1.3 of Chapter 2).

The goal of the overall system is to perform a task on the distributed instance of the data without moving the raw data at each fusion client around. The task may be learning an AI model, using an AI model to make an inference, or compute some other metric on the distributed data. The general approach for performing the task is for each fusion client to calculate a local model, inference or metric and share the metrics with the fusion server. The fusion server would then combine or fuse the models, inferences or metrics provided by all of the fusion clients and provide the fused version back to them.

3.1   Federated Learning of Metrics

As a component step of training and using AI models, it would often be necessary to obtain some metrics that apply to all of the data that is distributed across different locations. These metrics would be calculated over the entire amount of the data but need to be done without moving any of the data around.

For the sake of illustration for this section, we assume that the data at each fusion client consists of a set of inputs $x_1, x_2, ...x_N$ and the output y. Each fusion client has several instances of this data and we would like to compute various metrics on this data across all the fusion clients. Let us assume that all of the metrics are numeric.

Some metrics that are easy to aggregate across the different fusion clients without moving any data around are attributes of any particular type of data, such as the total count, mean, minimum, maximum and standard distribution/variance of any x_i across different fusion clients. The calculations can be done across the different fusion clients without moving their data around. Each of the fusion clients can send its local attributes over to the fusion server, who can then aggregate them to get the overall estimate.

The aggregation process for some of the metrics may include sending it along with the total data count. When mean, standard distribution or variance need to be aggregated, each fusion client can send over the locally computed metric along with the total number of data points. The fusion server can average the metric with the weight to each fusion client metric, determined by the total number of data points along with them. For other metrics, like minimum or maximum, the minimum reported by all of the fusion clients is the aggregated minimum and the maximum reported by all of the other fusion clients makes up the aggregated maximum.

Aggregated distributions or statistics about the data that is distributed can also be computed without moving the data around. Let us assume that the fusion server needs to compute the overall distribution of the different x_1 across all the fusion clients. This would require two rounds of communication between the fusion clients and the fusion server. In the first round, the fusion client would send the range (i.e. maximum and minimum) of x_1 to the fusion server. The fusion server would determine the aggregate range, and also identify the number of buckets into which data has to be divided. Each local fusion client can then count how many instances it has within each of the ranges, and send them over to the fusion server. The fusion server can then aggregate the buckets across all of the fusion clients and find the aggregate distribution.

An example across two fusion clients is shown in Figure 3.2. Each fusion client has some data. In the first step, the fusion clients send the range of their data to the fusion server, which is 0-15 and 5-20, respectively. The fusion server computes the overall range as 0-20, and instructs each fusion client to divide it into 4 groups of equal size, namely 5 units. In the second step, each of the fusion clients computes the histogram of distribution on these groups, and sends it over to the fusion server. The fusion server can compute the overall histogram by summing the entries in the corresponding buckets.

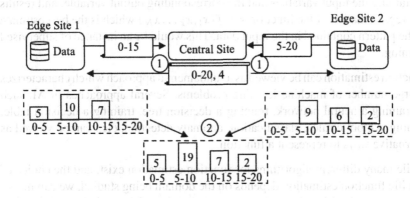

Figure 3.2: Federated statistical distribution calculation.

A similar process can be used across many fusion clients and to compute other type of metrics including percentiles, medians, or any other metrics. In general, any computation that can be broken into two components, a fusion client executable module and a fusion server executable module, can be implemented so that data does not need to move around. Each of the fusion clients invokes their fusion client executable modules to generate an intermediary output. The intermediary output can be sent to the fusion server which can process the information across all of the intermediary outputs from different fusion clients, and can generate the final output. This process can be repeated over in order to compute a complex set of functions without moving any data around.

This approach is a wide-area invocation of functions that are frequently used for parallel computation within a site. In general, any approach that can be implemented with a split data processing paradigm (e.g. map-reduce [33]) can be implemented in a federated manner. Specifically, any types of Key Performance Indicator (KPI) metrics used by businesses can be calculated in a federated manner using this approach.

3.2   Function Estimation

Estimating functions that fit on the data available at each fusion client is a generic type of AI model building which is required for a variety of business operations. In this section, we explore how that function can be estimated when data is distributed across several fusion clients with the assistance of a fusion server in the configuration illustrated in 3.1. The underlying technique for function estimation would be similar to that of federated averaging [30, 31].

Let us assume that the task of an AI model is to take in some N variables $x_1, x_2 \ldots x_N$, and to produce a value y. In other words, we assume that the process of machine learning takes an input training data consisting of many records, each containing the input variables and the corresponding output variable, and results in a representation of the function $y = f(x_1, x_2 \ldots x_N)$ which is the best estimate of the pattern contained in the input data. This would be an instance of supervised learning.

Function estimation can be viewed as a fundamental approach which characterizes a large number of machine learning problems. Several approaches in AI, such as training a neural network, training a decision tree, training a decision table, leaning symbolic rules, classification, anomaly detection, etc., can be viewed as alternative ways to represent a function.

While many different algorithms for function estimation exist, and the choice of specific function estimation depends on the domain being studied, we can model the general task of function estimation as a mapping of the available training data

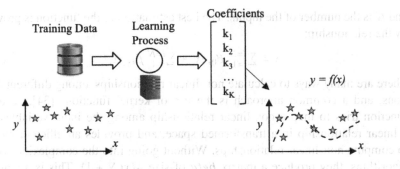

Figure 3.3: AI model building as function estimation.

set to a set of K numbers. The value and semantics of these K numbers depends on the algorithm used for function estimation.

The mapping of a set of training data to a set of K numbers is described visually in Figure 3.3. The training data is assumed to have only one input x which is to be mapped to an output y. The restriction to one is purely for illustration since 2-dimensional charts are easy to explain. The training data can be visualized as per the graph shown on the left side of Figure 3.3, where each data point is marked with a star. However, each of the data points is independent, and the patterns in the data are not captured in any way. After the training process, a set of K numbers $K_1 \ldots K_K$ are generated, which are a way to represent to a best fit estimate of the functional relationship between x and y. That function is shown as the dashed curve in the graph to the right of the figure.

There are many different algorithms that can be used to compute the K numbers that represent the function. As an example, linear regression [5] is a very common algorithm used within AI model learning approaches. The task of linear regression is to find the best fit linear relationship among the input variables that can predict the output variable. In this representation, the task of learning a function with inputs $x_1, x_2 \ldots x_N$ and output y results in a function representation where $K = N + 1$, and the machine learning calculates $N + 1$ numbers $\alpha_0, \alpha_1 \ldots \alpha_N$ which provides the best guess for a relationship between the output y and the inputs by means of the linear equation

$$y = \alpha_0 + \sum_{i=1}^{i=N} \alpha_i x_i$$

While the linear regression captures only a linear relationship among the input and output, there are many machine learning algorithms which capture non-linear relationships as well. While there are many algorithms to compute non-linear relationships, they eventually result in the creation of a matrix *beta* of size $Mx(N+1)$ where M is the highest power of any input variable that is considered, and N is the number of the inputs. The best estimation of the function is provided by the relationship:

$$y = \sum_{i=1}^{i=N} \beta_{0,i} x_i + \sum_{i=1}^{i=N} \beta_{i,j} x_i^j$$

There are many ways to calculate non-linear relationships among different functions, and a common approach is the use of kernel functions [34]. A kernel function tries to map a non-linear relationship among the input variables into a linear relationship in a transformed space, and provides an efficient method to compute non-linear relationships. Without going into the complexities of the algorithms, they produce a matrix *beta* of size $Mx(N + 1)$. This is a compact way of representing $Mx(N + 1)$ numbers. In other words, K equals $Mx(N + 1)$.

Yet another approach for function estimation uses a set of basis functions. The function that is being modeled is assumed to be a combination of a set of predefined basis functions. Basis functions are a set of predefined independent functions whose combination covers all functions of interest in a specific domain. The estimated function would be computed as a linear combination of the basis functions. The basis functions $b_1 \ldots b_K$ are functions defined on the inputs $x_1 \ldots x_N$ and their linear combinations would cover all the other functions of interest for the machine learning problem. The goal is to come up with the K coefficients $\gamma_1 \ldots \gamma_K$ which provides the best estimate for the relationship:

$$y = \sum_{i=1}^{i=N} \gamma_i b_i (x_1 \ldots x_N)$$

Common examples of the basis approach include wavelet analysis methods [35] and the Gaussian Mixture models [36].

Regardless of the approach used for function estimation, the net result is the calculation of the K numbers (or parameters) required to estimate the right function. Almost all machine learning algorithms result in some K numbers which characterize the function that is learned to capture the patterns contained in the data. When the model is a neural network, these parameters are the weights of the neural network. When the model is a decision tree, these parameters provide the number of nodes and the thresholds for branching at the different nodes of the decision tree.

3.3   Federated Learning for Function Estimation

In order to deal with the challenge of distributed data, a federated learning approach would estimate the function independently at each of the fusion clients, and subsequently aggregate all of the functions at the fusion server site.

The abstract process for fusion can be seen in Figure 3.4 for three fusion clients. Each of the fusion clients has some local data. The local data of the fusion client

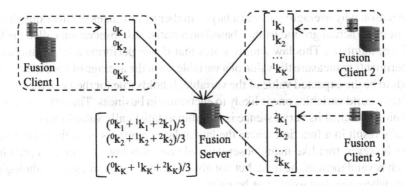

Figure 3.4: Federated learning for function estimation.

is used to compute the K numbers that can estimate the model on data local to the fusion client.

These numbers are sent to the fusion server, which averages them together. In enterprise computing, there are only a modest number of fusion clients working with an individual fusion server, and instead of just averaging the model once, each of the fusion clients computes the K numbers locally in multiple iterations. In each iteration, a random subset of the data present at the fusion client is selected, and those K numbers are calculated and sent over to the fusion server, which averages those values. These iterations are repeated until the average has been calculated a reasonably large number of times.

Let us understand the basic rationale for why the averaging over multiple iterations would result in a function that faithfully captures the patterns in the data throughout the set of fusion clients. For the task of function estimation, the assumption is that there is a ground truth function which can capture the patterns contained in the overall data. The subset of data that is contained at each of the fusion clients is a different portion of the overall data. The machine learning algorithm at each fusion client is analyzing a slightly different sample of the data, so slightly different functions will be learned. However, these functions are different variations of the same ground truth. If one were to average across hundreds of such variations, chances are very high that the average resulting function would be the same as the ground truth.

In an enterprise setting, we would have a few fusion clients, and learning the function from all of the data and then averaging them together may not provide a good approximation of the ground truth. However, if we were able to make subsets of data from each fusion client, have each subset be of a reasonable size to get a decent approximation of the ground truth, and then average these approximations hundreds or thousands of times, we are likely to get to the ground truth.

The reason why averaging across a large number of estimates of functions should result in the actual ground truth is based on a statistical observation called the law of large numbers. This law simply states that if one performs a large number of experiments to measure the value of a variable, then the average of the experiments tends to be the expected value of the variable. It holds true for the large majority of the real-world variables one is likely to encounter in business. The task of machine learning, which at its very essence is learning a statistically valid function, should usually result in a function where the law of large numbers is likely to hold true. Note however, that like many observational laws, this law is more of a principle which is valid most of the time, but not always, that is, one can create pathological cases where this law would not be valid.

The application of the law of large numbers to the task of federated learning is shown in Figure 3.5. On the left hand side, several estimates of the function are made, and these estimates are shown in the different grey lines shown in the graph. The dark line shows the ground truth. When a large number of estimates are averaged together, the resulting average would tend to come closer to the actual ground truth.

An illustrative example for using the averaging process for two fusion clients is shown in Figure 3.6. The figure consists of four graphs, each plotting one input variable x and the output variable y. The graph at the top left shows the total amount of training data that is available, where the stars mark the data that is present at fusion client A, and the circles mark the data that is present at fusion client B. Taken together, these stars mark the ground truth that is shown in the bottom left graph. That graph would be the function that would be learned if all the data were present at a single location.

If the two fusion clients learned their functions independently using the machine learning algorithm of choice, they would learn somewhat different functions. The functions learned by client A and client B, respectively, are shown in the two graphs on the right. The grey solid line shows the function that would be the actual ground truth, while the dashed black line shows the function each fusion

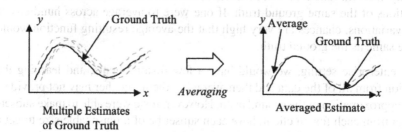

Figure 3.5: Averaging approach for federated learning.

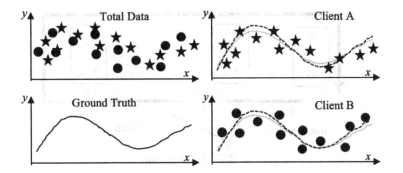

Figure 3.6: Federated learning for function estimation.

client would learn on their own. If one could average those two functions, the deviations are likely to cancel out and the average would become the same as the ground truth on the bottom left hand side.

Averaging over a small number of fusion clients (just 2 as shown in the figure) is not sufficient to get the law of large numbers to kick in and increase the likelihood that the errors are canceled out. However, if a random selection of a subset of data at each of the fusion clients is taken so that the function is learned several hundred times, the averaging process has a very good chance of eliminating the errors and coming close to the ground truth.

The averaging of the functions across a large number of sub-samples may be too time-consuming to be practical in some cases. However, it holds independently of the type of function that is being learned in the machine learning process. There may be some optimization that can be made to accelerate the learning process and the fusion can be done faster. One such optimization is to try to incrementally improve the parameters that are learned, which can be best illustrated in how neural networks are trained.

3.4 Federated Learning for Neural Networks

A neural network is a very popular approach for representing a general function that is extracted from a set of training data. The parameters that characterize a neural network are the weights assigned to the different neurons in the network, and are basically the K parameters described in Section 3.3. Instead of using random sub samples and averaging the resulting weights together, a faster technique that can be used in the learning process is that of iterative adjustment of model weights.

This process for training a neural network model is shown in Figure 3.7. The process starts with a set of initial weights that are randomly guessed or assigned.

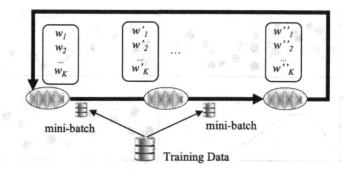

Figure 3.7: Iterative training of neural networks.

The entire training data is divided into multiple subsets, which are called a mini-batch. The weights of the neural network are used to find how different the predictions of the network using that weight are from the real output. Then, the weights are adjusted so that the error can be reduced. There are several algorithms for weight adjustment but going into the details of those algorithms is not important at this level of explanation. With the adjusted weights, the process is repeated for another mini-batch of the data. The process may be repeated by making multiple passes over the data, and each pass would reduce the error between the prediction and the actual value. The weights that are calculated at the end of this process define the trained neural network to be used for inference.

The variation can be seen as a distributed implementation of the approach described in Figure 3.6. The overall data is distributed across many different fusion clients, e.g. it is divided into three fusion clients, as shown in Figure 3.8. Each of the fusion clients would go through the process of training the weights iteratively, as shown in Figure 3.7. However, after each of the fusion clients has gone through one mini-batch training independently, they would send the model weights to the fusion server and the weights calculated by all the fusion clients would be averaged together. Each of the fusion clients would then start the next iteration of their mini-batches with the averaged weights.

In effect, the mini-batch that would have been used for a single fusion client has been replaced with mini-batches that are aggregated over all the fusion clients that are involved in the training process. Since the logic for training these models is the same as that of training the data at a single location, the model fusion process eventually results in the same function that the centralized approach would generate. As long as all the mini-batches are processed in some order, and depending on the learning algorithm, all the data has been examined one or more times, the learning process should approximately result in the function that would have been learned by collecting all the data to a central location.

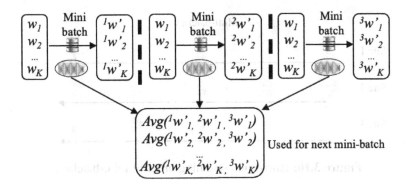

Figure 3.8: Federated averaging for neural networks.

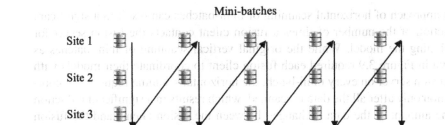

Figure 3.9: Vertical scanning order of mini-batches.

The manner in which the different mini-batches of the data at different fusion clients is scanned according to the algorithm described in Figure 3.7 is illustrated in Figure 3.9. If the data at each fusion client is divided into multiple mini-batches, the first mini-batch of each of the fusion clients is processed first, then the federated learning algorithm scans through the second mini-batches at each fusion client and progresses in this manner until all of the data is scanned. The data can be scanned multiple times if needed.

It is not necessarily the only order in which we can scan the mini-batches to cover the entire data at all fusion clients. Instead of scanning the array of mini-batches in a vertical sequence, one can scan the mini-batches in a horizontal sequence. The horizontal sequence of scanning mini-batches is shown in Figure 3.10. In this process, all the mini-batches at a single fusion client are scanned first, then the mini-batches at the second fusion client are scanned and so on.

In practical terms, horizontal scanning simply means that the entire neural network is trained at one fusion client, then passed to the second fusion client for training, and it passes sequentially through all of the fusion clients. After making several

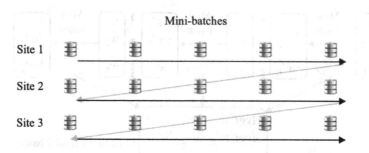

Figure 3.10: Horizontal scanning order of mini-batches.

passes through the set of all the fusion clients, the neural network training passes results in an estimate of the system that is comparable to that of training centrally.

This approach of horizontal scanning of mini-batches can result in a significant reduction of the number of times a fusion client contacts the fusion server for combining the model. While the original vertical scanning of mini-batches as shown in Figure 3.9 required each fusion client to coordinate their models with the fusion server on every mini-batch, the horizontal scanning requires the coordination only after all the data is scanned, which results in a significant reduction of the amount of the data exchanged between the fusion clients and the fusion server.

An alternative approach for horizontal scanning would be to have each fusion client train all of their models in parallel, so that each of the horizontal layer of mini-batches are traversed in parallel. When the fusion clients have finished training, they can average their neural network weights. The process would have to be repeated a few times so that the law of large numbers can kick in, and bring the overall trained neural network to the one that would have trained in a centralized manner.

Horizontal and vertical scanning are not the only options to traverse the different mini-batches of data stored at each of the fusion client. The fusion clients may choose to train over a different number of mini-batches, and synchronize at different times. The fusion server can maintain the averaged weights of the neural network, and let each fusion client connect with it and update with the results of its current mini-batch at different intervals. The averaging of the weights can be done with different level of importance given to different fusion clients, e.g. if a fusion client has twice the data of the other fusion clients, its model weights are counted twice when averaging over the weights provided by all of the fusion clients.

These different approaches for traversing the different mini-batches give rise to different types of federated learning, which are described in various papers such as [25, 30, 37, 38]. There are differences in the network bandwidth consumption, convergence time, and load on the fusion server imposed by the different algorithms, but they all operate in the overall approach described in this section.

3.5 Federation of Miscellaneous Models

The averaging of parameters can work across a wide variety of AI models if their result is a set of numbers where it would make sense to average the resulting parameters of the model. However, some types of models are not represented in a way that would permit a simple averaging.

Looking at the broad categories of AI models discussed in Section 1.5 of Chapter 1, we have already discussed functional models and neural networks which are good candidates for repeated averaging across different fusion clients. Some of the other models are also represented as a set of parameters which represent a mathematical functions. In those cases, federated averaging would work well to fuse these types of models together. Specifically, Support Vector Machines can be averaged together [39] in this manner. AI models which are represented as matrices, such as Principal Component Analysis [14], can also be federated together using an averaging approach [40].

Some models are not straightforward to average together if they are fully trained, e.g. decision trees. A decision tree has several branching points and, if the fusion clients are training their decision trees independently, the decision trees can not be combined in a straightforward manner. However, the existing decision tree training algorithms create their decision trees incrementally. At each stage of training a decision tree, the tree is partially built and a new node with a test needs to be selected. The test requires the selection of one of the input variables $x_1 \dots x_N$ to be used for the test and a threshold for the variable. This approach of growing a tree top down [41] would select the variable and threshold based on an evaluation of the value to be gained by the variable and the threshold. If all nodes select their variables, the majority vote can be taken among the different fusion clients to determine the next variable to be used for growing the decision tree. Once the variable is selected, the best threshold can be suggested by each of the fusion clients and a weighted average to select the suitable threshold determined by the fusion server. The individual step in each of these decision points for incremental growth would follow the federated approach for computing a metric as described in Section 3.1.

Some other models, such as decision tables and decision rules, are easy to federate. All the entries in a decision table can be put into a single larger table and analyzed for conflict among the areas where multiple entries correspond. The inconsistencies can then be identified using known techniques [42] and eliminated either by selecting in favor of what the majority of fusion clients are predicting, or by human inspection.

3.5.1 *Distributed Incremental Learning*

Some of the training algorithms for different types of models are incremental, in the sense that they follow the paradigm for learning a model and then improve that model using additional data. The approaches of training on mini-batches to improve the neural network weights is an example of an incremental process where an existing neural network is improved upon. Just as neural networks can be trained upon incrementally, other types of models can also be trained incrementally. Algorithms that perform incremental learning are known for a variety of function estimations that try to optimize some attributes [43], rule sets [44], decision tables [45], decision trees [46] and support vector machines [47].

If a training algorithm is incremental, it provides an alternative approach to perform federated learning. If each of the fusion clients have the ability to train the same model incrementally, the fusion server can arrange to have the fusion clients train a model in a sequential manner. A random order of training among different clients can be determined, the model trained on the first fusion client on the site that then uploads the model to the fusion server. The fusion server sends the model to the second client in the order, who trains the model further using an incremental algorithm. The model is sent next to the third client in the order. This process is repeated until the model has been trained through all of the sites. The process can be repeated until a few passes have been made through all of the sites. Repeating the training process through all the data repeatedly helps some models to improve themselves.

This process of leveraging the clients to train the models in a sequence using incremental learning is shown in Figure 3.11. Three fusion clients are shown along with the fusion server. The fusion server selects the random order to run

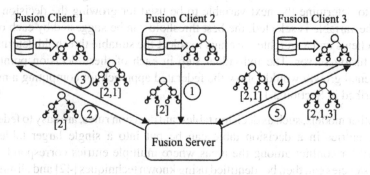

Figure 3.11: Distributed incremental learning.

a model through the fusion clients in the order of [2, 1, 3]. The fusion client 2 would use its local data to train the first instance of the model which is sent in step 1 to the fusion server after the training is completed. The fusion server then sends it to fusion client 1 which trains the model incrementally using its data. The resulting model which is now trained from data from fusion clients 2 and 1 is sent back to the fusion server. The fusion server would then send the model over to the fusion client 3, who will use its incremental training algorithm to update the model which now has all the data from three sites used to train it.

One drawback of the process shown in Figure 3.11 is that only one fusion client is actively training the model at any given time. This does not use the significant amount of parallelism that is available within the system and does not use the computation capacity available at the different fusion clients efficiently. To address that limitation, the fusion server can orchestrate the training processes in the clients so that all there are N models being trained in parallel, each model starting with the data from one of the N fusion clients. A random ordering can be determined by the fusion server so that each of the fusion clients is training a model in parallel instead of executing them sequentially, as shown in Figure 3.11. Once a random order is determined, the fusion server can create a rotated schedule to exchange the models that are trained with different sites.

Using this approach, suppose the fusion server determines the permutation order to be [2, 1, 3]. It can shift the order to create two more schedules which are [1, 3, 2] and [3, 1, 2]. The parallel training of the models happens in the manner shown in Table 3.1. We label the models A, B and C which will be trained originally at fusion clients 1, 2 and 3 as shown in first row of this table. After this step of training is completed, model A is trained at client 3, model B at client 1 and model C at client 2. After the completion of this second step, at the third step, model A is trained at client 2, model B at client 3 and model C at client 1. At each step, each model is being trained at one of the fusion clients. At the end of the process, the three models have each been trained on all three of the data sources in a different order. The cycle shown in the table can be repeated if needed.

Given the stochastic nature of training of AI models, the three models will have some differences and not give exactly identical answers to all inputs. However, at the end of the training process, these models can be put together into an

Table 3.1: Parallel training of models with incremental training algorithm.

Step	Model A	Model B	Model C
1	Client 1	Client 2	Client 3
2	Client 3	Client 1	Client 2
3	Client 2	Client 3	Client 1

ensemble [48] and used by all of the fusion clients during the infer phase of the Learn→Infer→Act cycle.

3.6 Assumptions in Naive Federated Learning

Although the different approaches for federated learning described in Section 3.4 seem to be good from the perspective of training a model from data distributed across many different fusion clients, these approaches are all reliant on some common assumptions. These assumptions need not be satisfied in real-life enterprise contexts or may cause some challenges in the logistics and operation of the same. In this section we examine some of those assumptions.

The algorithms in this chapter assume that the training data used at the different fusion clients are consistent across all sites and can be used to train a common model. If the data is not consistent across sites, i.e. different sites have different types of inputs, or inputs are not scaled consistently, the models parameters can not be averaged.

In practice, the factors which prevent data movement are also the ones that drive a significant amount of inconsistency among the different sites. If data is not allowed to move across different fusion client sites, that data is governed by different people or subject to different guidelines. In some cases, that data may have been collected independently. As a result, the data would likely not be consistent across the different fusion client sites.

The various mismatches that exist among different fusion clients data sets, the problems they can cause and approaches to address them are explored in more depth in Chapter 4.

The Naive algorithms assume that the data that is present is not skewed. Data skew happens when different fusion clients collect data with very different characteristics. As an example, one fusion client may have collected a lot of data that captures the purchase behavior of people in an urban environment. Another fusion client may have collected data that may be representative of purchase behavior of people in a rural environment. We can learn neural networks or functions to predict the purchasing characteristics of the population by averaging the behaviors across both of the environments, but simply averaging them may not be the right way to merge the two models. These are two very different functions that are being learned, and the assumptions underlying federated averaging are just not valid in this context.

The assumption implicit in the naive federated learning approaches is that all of the fusion clients are learning the same function. If the data at all of the fusion clients has the same common function that is being estimated, the averaging of the parameters in any order would eventually lead to the same function. Data skew will lead to a violation of this assumption. The function being learned at different fusion clients may be different if they maintain their data differently, and have collected data about different types of environment.

The challenges caused by data skew and approaches to address them are described in Chapter 5.

An implicit assumption in the algorithms presented in this section is that the fusion clients and fusion servers trust eachother to behave in a proper manner. The fusion server and fusion clients are all working properly and all models being exchanged are maintained according to policies and guidelines acceptable to everyone.

When federated learning is happening in environments which involve more than one organization, trust among different organizations may not be absolute. In those cases, mechanisms may need to ensure that issues of limited trust be addressed and that the infrastructure and algorithms for federated learning work in these environments.

The scenarios for working across different trust boundaries and approaches to address them will be discussed further in Chapter 6.

Another assumption in the algorithms presented in this chapter is that the all the fusion sites are active at the same time and are training in a synchronized lock-step manner. The vertical scanning of mini-batches described in Figure 3.9 can only be done if all fusion clients are active at the same time. However, in many real-world scenarios, a much more asynchronous approach for training models may be needed. Models from different sites may be available at different times.

An implication of the synchronization assumption made in the algorithms is that the models at each of the fusion clients share a common architecture. Each of the fusion clients are learning some K parameters but these K parameters have to be the same at each fusion client, and have the same semantics. If we are training a neural network with some L layers of neurons and some N neurons in each of the layers then $K = L.N$. However, it does mean that each fusion client must be training a neural network with the same number of layers and the same number of neurons in each of the layers. This means that all fusion clients have to agree to a common architecture for models before training.

To understand the rationale for requiring a common architecture, let us examine two different neural networks that can be trained to do the same task, e.g. learn from the attributes of a loan application as to whether the loan is in one of four categories of risk. Risk, here, is an estimate of the likelihood that the borrower would not be able to pay back the loan. Many different kinds of neural networks can be trained to do the same, and two simple examples of neural networks that can classify the input attributes and output a decision into one of four categories are

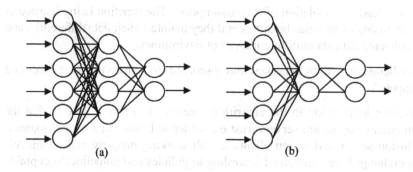

Figure 3.12: Two alternative neural networks for same task.

shown in in Figure 3.12. Note that these simple examples are only for illustration, real-world neural networks are going to have many more layers and a larger number of attributes and neurons per layer.

In the neural network (a) shown in the figure, there are 7 input nodes, there is a middle layer of 4 nodes and there is an output layer of 2 nodes. In the network (b), the number of input nodes and output nodes is the same, but the middle tier only has 2 nodes. Assuming a weight is associated with each of the input links (or neurons) for each node, the neural network (a) will have 4*7 + 4*2 = 36 weights, while the neural network (b) will have 2*7 + 2*2 or 18 weights. Therefore, the neural network (a) will be characterized by 36 numbers, whereas neural network (b) will be characterized by 18 numbers. However, both of these two networks can be trained to do the exact same task, mapping from inputs with 7 attributes to an output that takes four possible values.

If during the federated learning process among 2 fusion clients, one of the client decides to use the neural network (a) to train on its data, and the other fusion client decides to use the neural network (b) to train its data, those two neural networks weights can not be combined together. In order to do the averaging of parameters and to fuse the models, networks at both fusion clients ought to have the same architecture.

In a consumer federated learning context, where the models are trained by different instances of the same mobile application, such an assumption is not unreasonable. Each mobile phone would use its data to train a model using an architecture that was defined by the developer of the mobile application. However, in an enterprise federated learning context, data is maintained in several locations which may be under different administrative or regulatory jurisdictions. This makes the coordination task of ensuring that each fusion client has the same model architecture relatively more complex. If the company has a part that is obtained via an acquisition, the acquired company may be using a different type of model to make its decisions. Those pre-trained models cannot be used for straightforward federated learning, and somehow all parts of the company have to agree to the same common architecture. This agreement on the common model can be very difficult depending on the structure of the company.

The second assumption that is made in the algorithms is that all fusion clients should be training at the same time. The approaches for federated learning require model weights be learned and trained at all the fusion clients at the same time. This would require that the fusion clients start training at or about the same time, periodically exchange the model weights and proceed with the training part in a synchronized manner. Some of the synchronization issues can be addressed by means of fusion clients coming and updating the aggregated model weights maintained at the server. Given the fact that different enterprise silos containing distributed data are probably managed independently, such a coordination would be challenging. In some cases, one can assume that a trigger from the fusion server can start the training process. This trigger can be created when the fusion clients trust each other, or at least the triggering fusion client. However, in settings like a consortium or a military coalition, such a coordinated trigger process may not be feasible.

It would not be unreasonable for different semi-independent or independent parts of the organization to train their models at different times. Even if there is a mandatory or voluntary conformance with a common model architecture, this means that models are trained at different times. Combining models trained independently by averaging once does not work well since the law of large numbers would not be applicable. This makes the task of training at different times very challenging for the system.

These issues are explored in more detail in Chapter 7.

Data partitioning arises when data about the same individuals is collected for different purposes, so that an entity monitored at one site has features and attributes that are very different than the features and attributes that are present for that entity at another site. If one looks at the various companies that are active within a given area, they are providing services to the same population. However, given the nature of their business they would collect very different types of information about the people in their service area.

Partitioned data breaks the assumption that the models in the different fusion clients are the same. When different data attributes are present, different models will be created. These models can not be averaged together blindly, but it may make sense to use the different models together, if possible.

Different approaches can be used to combine the insights from the different models, including approaches that can convert them to a consistent model and approaches that simply use the inconsistent models during the inference phase. These approaches are discussed in more detail in Chapter 8.

3.7 Summary

In this chapter, we have reviewed the basic approach for federated learning using iterative averaging. These approaches work when the different fusion clients are trying to build a model together, have data that is distributed relatively uniformly among them, trust each other and can synchronize with each other in order to train a model. In real-world environments, many of these assumptions will not be satisfied. In the next few chapters, we will look at some of the approaches to modify and adapt the naive approaches so that the challenges in enterprise federated learning can be addressed adequately.

Chapter 4

Addressing Data Mismatch Issues in Federated AI

One of the major challenges for the federated AI is the fact that the data sets maintained within different physical locations are likely to be in different and inconsistent formats, have a varying quality and use different terms for the same concept. In order to train a model using federated learning, or to use the model using federated inference, all of the data at all the fusion clients must be converted to a single consistent format. In this chapter, we look at some of the techniques that can be used to attain this goal. Unless these mismatches in data at different fusion clients are resolved, federated AI can not be used meaningfully in either the learn or the infer phase of the Learn→Infer→Act Cycle.

In general, data used for machine learning at any fusion client can be either featured or raw (i.e. featureless). The raw data typically consists of unstructured text, images, video, sound, speech, sensor readings, etc., which are typically digital representation of continuous signals or natural language documents. Featured data is usually information that consists of several clearly delineated features, e.g. data for bank loans, health management records, retail sales transactions, etc. Featured data is typically represented in tabular format such as spreadsheets or relational databases, or in structured text format. Some examples of structured text format are javascript object notation or JSON [49] and extended markup language or XML [50].

During machine learning, the initial step is the conversion of raw or featureless data to a set of features, which are subsequently used to train the AI model.

In some cases, feature extraction may be done explicitly, e.g. see [51] for some techniques for extracting features from images. In other cases, specifically when using deep neural networks, the initial few layers of the neural network extract the features automatically as part of the training process. In those cases, the neural network itself can be viewed as consisting of two sub-networks, one doing the feature extraction and the other converting the features to the desired output.

Sometimes, the training data contains the output that results from a combination of features. Those training data sets are called labeled, since they contain the right labels associated with each of the data sets in the training input. Some learning approaches only work with labeled data while other learning approaches do not require the data to be labeled.

Data mismatch across fusion clients can take many forms. The following are some of the common data mismatch issues:

- **Format differences**: The data at different fusion clients may be in different formats. In order to create a combined model, these differences in formats need to be reconciled. When the data is raw or featureless, this difference may manifest itself as data being represented in different formats. When the data is featured, some features may be missing in the data at some of the fusion clients. Features may also be named differently across the different fusion clients, or be recorded in a different manner. As an example, a feature called Name may be recorded following 'FirstName Surname' convention at one fusion client, while it may be recorded as 'Surname, FirstName' at another fusion client. These differences need to be reconciled before the process of machine learning can be initiated.

- **Value Differences**: When the input data is featured, the values that different features take at different fusion clients may not necessarily be the same. One fusion client may have data that assesses the risk of loans with color codes like yellow, blue and green, while another fusion client may have the risk assessed numerically as 1, 2, and 3.

  One specific problem of value difference that is common across both featured and raw data arises when data is labeled, and the values assigned to the labels are different at different fusion clients. For example, if image data has been collected, a manufacturing site in France may have labeled images of different manufacturing products in French, while the American manufacturing plant of the same firm may have the labels in English. A consistent labeling scheme needs to be developed across all of the fusion clients in order to create a meaningful model.

- **Quality Differences**: Different fusion clients may have data with different quality. In order to build a good model, data from fusion clients that may have worse quality than average needs to be eliminated. It would be useful to only use data that has a minimum level of quality for building the model. In order to deal with this aspect, it is important to have a definition of data quality, and an approach to compute data quality.

■ **Partitioning of data**: The data that is available at different locations may be partitioned in different ways. Partitioning of data can be explained best in terms of featured data with labels. Consider the featured data as being stored in tables where the columns are the name of the features with the last column being the output, and that the rows are sorted according to the labels. The partitioning can then be defined as being either horizontal or vertical. In vertical partitioning of data, some of the features may be missing from the data at some fusion clients, i.e. some of the columns of the tabular representation are missing. In the horizontal partitioning of the data, some of the labels may be missing from data at different fusion clients, i.e. a block of rows with a given label is missing from the data at some fusion clients. Both types of partitioning introduce challenges in the fusion of AI models that are trained at different fusion clients.

In the different sections of this chapter, we explore approaches for resolving these differences in order to create models that can be combined with each other at the fusion server. Partitioning Issues are addressed in Chapters 5 and 8, respectively.

4.1   Converting to Common Input Format

The approaches to convert to a common format are different depending on the type of data that is available for training. If the data type is raw, conversion is the task of changing the format of the data, e.g. convert an image from the JPEG format to the TIFF format. If the data type is featured, conversion may require changing names of features, or adjusting values of the features. Furthermore, if there is a normalization to be performed on the features, this normalization ought to be done in a consistent manner across all the fusion clients.

4.1.1   *Raw Data Types*

Let us consider the instance when different fusion clients have raw data (i.e. without explicit features), and the data is maintained in different formats at different fusion clients. Although the data may be of the same type, there are many formats to store each of the data types.

Some of the common data types and the formats in which they can be stored at different fusion clients are shown in Table 4.1. Given the wide set of choices for the same data types, it is not unreasonable to expect that some of the fusion clients may have chosen different formats.

Table 4.1: Common formats for some raw data types.

Data Type	Formats
Images	jpeg, tiff, eps, png, svg, bitmap
videos	quicktime movie (.mov), windows media video (.wmv), mpeg 3(.mp3), flash video (.flv)
Sounds	wav, mp3, webm
Text Files	rich text format, microsoft word, pdf, plain text, open document (.odt)

Table 4.1 only lists a few types of raw data types and a few common formats for representing each of those data types. There are many more examples for each of the raw data types, e.g. many other formats for images beyond the six listed in the table are found across different sites. However, the problem for creating a federated AI approach remains the same, different fusion clients may have raw data in many different formats. However, the creation of a common federated model requires the data to be in a single format across all of the fusion clients.

An approach to select the format is to pick the one in which the maximum amount of data is available, and forego the use of data that is in other formats. This approach avoids the task of format conversion. However, it may end up excluding a significant amount of available data. In machine learning, data is very valuable, and expanding upon the set of available training data is the most important reason driving federated learning. Not using available data is not a good choice if the goal is to build good AI models.

The conversion of data in other formats to the selected common format can be done by means of format conversion software. Many such format conversion software packages are available, some in open-source, some as proprietary software and some as Internet-based services. The availability of the software conversion capability to different fusion clients can be used to select the right common format to use for model building.

In order to determine the right format, each fusion client can create a conversion graph. A conversion graph is a graph whose nodes represent one of the available formats of data, and a directional edge connects two nodes if format conversion capability in the direction of the arrow is available at the fusion client. Each fusion client would have a different conversion graph determined by the availability of the software packages and available services to do the conversion to different fusion clients. This conversion graph, along with the total amount of data that is available in each format locally is reported by each fusion client to the fusion server. With the conversion graphs of all fusion clients, the fusion server can identify the format which will provide the best model across all of the fusion clients.

If data conversion costs are negligible, the right format to choose would be the one that results in the maximum amount of data being available for creating

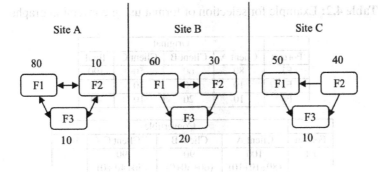

Figure 4.1: Example of conversion graphs.

the model. For each of the conversion graphs, a metric can be calculated as to what the total amount of data available at each fusion site would be based on the conversion graph. An illustrative example is shown in Figure 4.1. This assumes that there are three fusion clients and each has data in three different formats. The conversion graph which shows which of the formats can be converted to each other is shown in the Figure. Each site can also provide the original amount of data that is available in each of the formats at the site. In the figure, we can assume that number is provided in the thousands of samples that are available.

Once the conversion graphs for each of the fusion clients are available at the fusion server, it can determine how much data can become available by converting to each of the possible available formats. For the specific set of conversion graphs shown in Figure 4.1, the amount of data that can be processed at each fusion client is calculated in Table 4.2. Each row in the table shows the number of data samples at each fusion client which can be converted to the format in the first column. The numbers in the Original columns show the amount available at the fusion client in that format. The numbers in the Convertible columns show the amount of data that can be converted to the specified format at each of the fusion clients. The second row in the Convertible column shows the format from which the data is being converted, with the convention being F1+F2+F3 enumerating the data that can be converted from the specific format.

On examining the amount of data available originally in each of the formats, we can see that format F1 is most common with 190 million samples across all of the fusion clients, compared to F2 (80 million samples) or F3 (40 million samples). However, because of varying ability of different fusion clients to convert to different data formats, it is better to select format F3 for conversion. Selection of F3 allows for 300 million samples to be used for training the data, as compared to the most common format F1, whose selection only allows for 280 million samples to be used for training. The conversion graph concept enables the determination of the format to which most of the samples can be converted.

Table 4.2: Example for selection of format using conversion graphs.

Format	Original			
	Client A	Client B	Client C	Total
F1	80	60	50	190
F2	10	30	40	80
F3	10	20	10	40

Format	Convertible			
	Client A	Client B	Client C	Total
F1	100 (80+10+10)	90 (60+30+0)	90 (50+40 +0)	280
F2	10 (80+10+10)	30 (60+30+0)	40 (0+40+0)	80
F3	100 (80+10+10)	100 (60+30+20)	100 (50+40+10))	300

The discussions so far based on Table 4.2 assume that there is no cost for converting data from any format to another. However, in many cases there may be a cost associated with the conversion, specially if a web-based service is used. Any such conversion costs can be represented as the cost of traversing the links of the conversion graph. This would allow a weighted value of the cost involved in converting the data and the utility provided by each additional data format. This weighted matrix can then be used to compare the cost-utility trade-off for each of the formats, and the right format can then be selected.

Let us assume that the utility of using each additional data point is 0.1 units, while the cost of converting the items into different formats at various fusion clients is as shown in Figure 4.2. In that case, the costs associated with using each of the different formats and the utility of using a given format are as shown in Table 4.3.

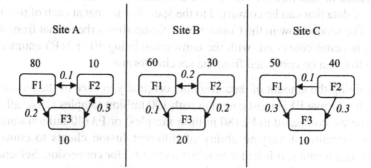

Figure 4.2: Conversion graphs with costs.

Table 4.3: Example for selection of format using costs.

Format	Client A Cost	Client B Cost	Client C Cost	Net Cost	Utility
F1	3 (0.1*10+0.2*10)	6 (0.2*30)	4 (0)	13	28
F2	11 (0.1*80+0.3*10)	12 (0.2*60)	0 (0)	23	23
F3	7 (0.2*80+0.3*10)	12 (0.1*60+0.2*30)	27 (0.3*50+0.3*40)	46	30

If we assume that the net value is the difference between cost and utility, we can see that format F1 is the best one to use.

Depending on the specific business context, a different function for combining cost and utility would be more appropriate. However, the analysis of format selection based on conversion graphs can be used in a similar manner.

4.1.2   *Featured Data*

In contrast to raw data, featured data consists of many different features and an output label. These features may be extracted from raw data, e.g. a sample of audio may be converted into features such as Mel-Frequency Cepstral Coefficients (MFCC) [52], which is a set of features commonly used in speech processing applications. In general, a featured data set can be viewed as the representation of a data in a table, where one of the columns is the output column and the other columns are the input columns.

When featured data is distributed across many fusion clients, each client at a different location, the features may be named differently at different fusion clients, i.e. they may use different names for the same column in their table of data. For the purpose of this section, we will assume that all the columns are present at all of the fusion clients. However, the columns may be named differently. Furthermore, some of the columns may be in a different format at different fusion sites.

Before the different fusions clients can start training a model, each of them need to convert the data to a common format, where the features and output labels are identical across all of the fusion clients.

In this section, approaches to handle differences in the naming of features, and different representations of the features are discussed. Approaches to handle missing features are discussed in Chapter 8.

An illustration of mismatch in data is shown in Figure 4.3 for a simplified case of a bank that is recording which of the loans it made were paid out completely and which were defaulted upon in several different countries. The data in all countries consists of the same features, i.e. name, date of birth, amount of loan and the class

Name	Date	Loan	Class
Dow, Joe	12/01/1994	300	Defaulted
Doe, Jay	01/13/1997	250	Paid
Doe, Sue	04/15/2000	300	Paid

American Format

European Format

Surname	Name	Date	Loan	Class
Dow	Joe	01.12.1994	0.3	Defaulted
Doe	Jay	13.01.1997	0.25	Paid
Doe	Sue	15.04.2000	0.3	Paid

Figure 4.3: Example of featured data mismatch.

of loan (which is either paid or defaulted). Fusion clients in all countries will be engaging in a federated learning exercise to create a common model. However, the fusion clients have access to data stored in two different formats, the American format and the European format.

Fusion client following American format store data in the name column as First Name, Second Name, the date column uses the format of month/day/year, and the loan counts the amount in thousands of dollars. For fusion clients following European format, the name is represented in two columns, Surname and Name (which indicates the proper name or first name), the date uses format dd.mm.yyyy (European convention), and the loan counts the amount in millions of dollars.

The two tables shown in Figure 4.3 contain the same data as it would appear at the two different fusion clients. The two tables look very different even though they represent the exact same data. Unless the data is converted to a consistent format, it would be difficult to make a meaningful model from the combination of data from both the sources.

Rule-based Conversion to Canonical Format The conversion process needs to be automated at each fusion client. An approach to perform the translation automatically would be to define a set of rules to translate and convert the entries in each column at each fusion client to a common format. The data scientist initiating the model learning process needs to define a set of translation rules for each of the original formats which is different than the common format.

For the particular example illustrated in Figure 4.3, let us assume that there are two fusion clients with data in the European format and three fusion clients with data in the American format. The data scientist has opted to use a canonical format which is different from both the American and the European format. To enable the use of data stored in the European format to convert the data into the selected format, a set of translation rules from European format to Canonical

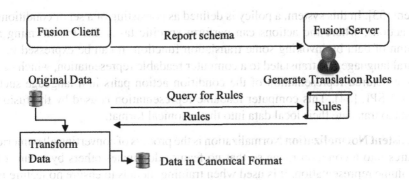

Figure 4.4: Interaction diagram for obtaining translation rules.

format would be defined. Another set of rules to transform from the American format to the Canonical format will also be needed. These rules will be stored at the fusion server, and each fusion client needs to retrieve the translation rules for its format from the server as an initial step.

The process can be implemented using the interaction diagram shown in Figure 4.4. In an interaction diagram, the sequence of messages flowing between different entities is shown. Initially, each of the fusion clients would report their schema (e.g. name of the local columns) to the fusion server. The data scientist at the fusion server would examine the reported schemas from all sites, determine the number of different formats, and define the translation policy for each of the formats. Once the translation policies are defined for each format, and each fusion client is associated with the format it is using, fusion clients can query the fusion server to retrieve the translation rules pertinent to them, and apply the rules in order to convert their data into common format.

Let us assume that the data scientist has defined the canonical format as consisting of the columns Surname, FirstName, Date, Loan amount in thousands, and Class. It examines the five fusion clients to determine that there are two different formats being used. For fusion clients using the first format (the American format), the translation rules would state that the Surname field is the part of Name field before the comma separator, and the FirstName field is the part of the Name field after the comma separator. The Date, Loan and Class fields remain unchanged. For fusion clients using the European format, the corresponding rules will state that the Surname field is unchanged, the FirstName field is the Name field, the Date field is obtained by changing the separator in the original data from a dot to a slash and reversing the month and date fields, and the Loan field is obtained by multiplying the original Loan field by a thousand.

The management and definition of these translation rules can be done in many different ways. We select one of the ways as the use of a policy management system [53]. In this system, a policy is defined as consisting of a set of conditions and actions, where the actions can perform specific tasks like transforming a column of data by invoking some translation function. It can be expressed in a natural language and translated to a computer readable representation, which can be a structured representation of the condition action pairs in a language such as CIM-SPL [54]. This computer readable representation is used by the fusion clients to translate their local data into the canonical format.

Consistent Normalization Normalization is the process of converting all numeric features into a common range so that no feature dominates others by means of its volume representation. It is used when training models to ensure no feature is being considered overly important when analyzing the data to train a model.

As an example of the need for normalization, let us consider the task of determining if a bank loan is risky based on the age of the borrower, their annual income level and the amount of loan being asked for. The age of the borrower would typically be a number between 20 and 100, while the annual income and amount of loan will be in several thousands. Thus, the latter is likely to overwhelm the parameters of any formula or AI models that are used to predict them. Furthermore, if instead of annual income, one were to use monthly income, the relative importance of the parameters would again change. Since large numbers tend to influence the model parameters disproportionately, keeping different parameters in different ranges would distort the models.

In order to get a better relative importance among different features, it is a recommended practice to scale all the numeric features to a common scale, typically a value between 0 and 1. Different approaches for this scaling are available and supported in various machine learning software packages [55].

For features that take on discrete values without an order, generally referred to as categorical features, the normalization step usually requires that they be converted to a set of binary values, with the number of distinct values of the feature determining how many such binary values are introduced. Common conventions include one binary valued variable for each distinct value of the feature, or one less than this number. Such conversion of categorical variables to binary values allows for more accurate models.

While normalization is relatively straightforward if all the data is at a single location, it can become problematic when data is stored at different locations. For numeric features, scaling approaches are based on the statistical properties of the data, such as the mean, max, or mean, depending on the type of scaling approach. These attributes need to be calculated across all of the fusion clients with the data. Without such coordination, different fusion clients may normalize the features differently. For example, if the convention is to scale data uniformly between 0 and 1 based on the minimum and maximum value of the feature, these values are likely to be different at each fusion client. A lack of coordination in scaling would mean that the same feature is being mapped differently by different fusion clients, leading to an inappropriate model.

Another type of scaling is to modify the parameter values so that they all are scaled to a distribution with a mean of zero and a standard deviation of 1. All fusion clients need to agree on the type of scaling to be used, and calculate the parameters of scaling consistently across all of them.

The type of interaction among fusion clients and fusion server that was used to reach a common canonical format described earlier can be reused to obtain a scheme for consistent normalization. Each fusion client reports the statistics of its local data (e.g. count, min, mean, max, etc.) to the fusion server. The fusion server can determine the corresponding values for all the data distributed across all the fusion clients, following the approaches described in Section 3.1 of Chapter 3. The normalization rules can then be defined, which would include the selection of the algorithm for normalization, along with the relevant aggregate statistics of the data.

Note that the fusion server can pick any normalization scheme which is consistent across all fusion clients, and it need not be based on the aggregate data properties. For example, it would be perfectly fine for the fusion server to determine the normalization rules based on the statistics it has for the first fusion client that contacts it. Every subsequent fusion client contacting the fusion server will be given normalization rules based on the statistics of the first fusion client.

As an example, let us consider fusion clients that are scaling all values linearly between the minimum and maximum value of a feature. There are four fusion clients with data distributed as shown in Table 4.4. In the table, only values for a single numeric feature are shown, but the concept would easily extend to multiple features.

With this example, without any coordination, the scaling formula for fusion client A for the feature shown in Table 4.4 would be $(x - 6)/22$, where x is the value of the feature for any of the 3,000 data points. The scaling formula for fusion client B would be $(x - 1)/22$, that for fusion client C would be $(x - 3)/43$, and that for fusion client D it would be $(x - 7)/20$. While each fusion client would be scaling

Table 4.4: Example for data normalization.

Fusion Client	Number of Data Points	Minimum Value	Maximum Value
Client A	3000	6	28
Client B	5000	1	30
Client C	9000	3	46
Client D	1000	7	20

their features to values between 0 and 1, the same original feature value would scale differently at different fusion clients. This would not lead to a correct AI model.

Before model learning starts, each of the fusion clients would need to send their aggregate statistics, the number of points, the minimum, the maximum, or the mean values of the feature to the fusion server. The fusion server can calculate the minimum among all fusion clients, as well as the maximum and send it back to each of the fusion clients. Then, each fusion client can scale according to the global minimum or maximum. In this case, the global minimum is 1 and the global maximum is 46, so the common scaling formula at each fusion client would be $(x - 1)/46$.

Do note that the alternative approach where the first fusion client to contact (e.g. fusion client A) the fusion server provides its value and the fusion server instructs all other clients to use the scaling formula of $(x - 6)/28$ based on the first fusion client to contact it. That would also be an acceptable scaling function across all clients, with the only issue being that it may scale some values to be negative (less than 0) and some to be over 1, if a fusion client has data that does not fall within the scaling range provided by the fusion server. However, since all fusion clients are using the same scaling formula, they would still be able to train a model that is accurately combining insights across all fusion clients.

For categorical values, the fusion server needs to determine the global set of values these categorical features can take. Some of the values may not be present in the data at some of the fusion clients. In order to define this mapping, a consistent naming scheme for each binary feature that is generated needs to be generated. If each fusion clients reports all of the unique values it has for each categorical variable, the fusion server can aggregate the overall set of unique values, and determine a unique set of values for everyone to use. With categorical values, it would be useful to have all fusion clients report their unique values before the overall set is generated.

Alternatively, the fusion server may decide to use a fixed number of binary variables for each category, and assign these values to different fusion clients as they contact the fusion server in sequence. One of these variables has to be reserved as a catch-all category in case the predetermined fixed number of categories is too small, and new values are discovered later.

An example for coordinating among categorical variables is shown in Table 4.5. It shows four fusion clients where a given feature takes three different values at each of the fusion clients. However, the values at each of the fusion clients are slightly different. Suppose the feature values are names of colors, the standard process of encoding the categorical variables as a One-Hot encoding can be followed.

Table 4.5: Example for categorical data normalization.

Fusion Client	Distinct Values
Client A	Blue, Red, Gold
Client B	Red, Blue, Yellow
Client C	Gold, Green, Yellow
Client D	Red, Blue, Green

Table 4.6: Categorical data encoding.

Fusion Client A		Fusion Client B		Fusion Client C		Fusion Client D	
Code	Value	Code	Value	Code	Value	Code	Value
[1, 0, 0]	Blue	[1, 0, 0]	Red	[1, 0, 0]	Gold	[1, 0, 0]	Red
[0, 1, 0]	Red	[0, 1, 0]	Blue	[0, 1, 0]	Green	[0, 1, 0]	Blue
[0, 0, 1]	Gold	[0, 0, 1]	Yellow	[0, 0, 1]	Yellow	[0, 0, 1]	Green

In One-Hot encoding, the categorical feature is represented as a binary vector whose length is the number of unique values the feature takes. Each value is represented as a array of ones and zeros. Each array has exactly one occurrence of a 1, i.e. all entries but one are zeros. The position in the array which is a 1 is different for each unique value of the feature. In the example shown in Table 4.5, the feature takes three distinct values at each of the fusion clients. Therefore, without coordination, each site would pick a binary array three units long and one possible encoding would be as shown in Table 4.6. As apparent from the table, the same value can get coded differently at different fusion clients. Furthermore, same encoding at different fusion clients may represent different values. While there is one-to-one correspondence between the categorical values and the encoded representation at any fusion client, this correspondence is lost when examining data across all the fusion clients. Combining the data from each of the fusion clients with this inconsistent encoding is going to result in a strange and inconsistent model.

If the fusion server is used for coordination of the categorical values, then the fusion server can determine that there are 5 distinct values: Blue, Red, Gold, Yellow and Green. This will allow the fusion server to recommend a One-Hot encoding that is 5 bit-long for the feature value. The net result would be a consistent mapping among the different fusion clients. The net resulting encoding is as shown in Table 4.7.

The synchronization for normalization does not require exchanging raw data. Instead, only summarized information about the data present at each fusion client needs to be exchanged. The process ensures training data is pre-processed and prepared in a consistent manner across all the fusion clients.

Table 4.7: Categorical data encoding with coordination.

Code	Value
[1, 0, 0, 0, 0]	Blue
[0, 1, 0, 0, 0]	Red
[0, 0, 1, 0, 0]	Gold
[0, 0, 0, 1, 0]	Yellow
[0, 0, 0, 0, 1]	Green

4.2   Resolving Value Conflicts

Value conflict is an issue that arises for categorical features, including the output labels which are categorical in many situations. If fusion client A is referring to the quality of its loans as green, yellow and red while fusion client B is referring to the quality of its loans as *vert*, *jaune* and *rouge* (French names for the same colors), we have a value conflict.

For featured input, value conflicts can arise in any feature (i.e. any column in the tabular representation). For raw input, value conflicts can arise in the output label.

To illustrate the problem, let us consider a classical problem of image classification. If images are maintained at different fusion clients independently or under different conventions, it is likely that the images have been labeled differently. As an example, let us consider the three fusion clients which are shown in Figure 4.5. The same type of images are listed as Autos at fusion client A, Cars at fusion client B and Vehicles at fusion client C. Similarly, what is labeled as flowers at fusion client A is called Plants at fusion client B and Blossoms at fusion client C. Each fusion client uses only two labels to describe the images it has captured, but the images are named and categorized very differently.

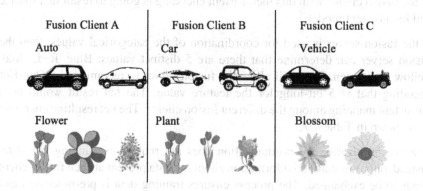

Figure 4.5: Example illustrating value conflicts.

If the different labels at different fusion clients are not reconciled, the model that is trained with inconsistent labels would have a poor quality. The same type of data would be defined with different labels and the model would be trying to learn differences among two groups that are essentially the same, e.g. the first two images at fusion client A and fusion client B are the same, but labeled differently. The challenge, of course, is to reconcile the conflict among all of these labels without requiring the data to be moved out of the fusion clients.

In the next few subsections, we look at the approaches to address the value conflicts and to resolve them.

4.2.1 *Committee Approach to Reconciliation*

The first and preferred approach would be to implement manual reconciliation of the different labels. The approach of manual reconciliation would be to define a set of translation labels for each of the fusion clients which determine how the labels at each client ought to be transformed. We can imagine a committee consisting of one person representing each fusion client sitting together over a web-conference and making a consensus decision on what the common labels ought to be and how the labels at each site ought to be mapped to that consistent label. This process is illustrated figuratively in Figure 4.6. For the example shown in Figure 4.5, one can decide that the final labels are Car and Flower, and the label remapping rules can be defined as the following:

- If fusion client is A and label is 'Auto', change label to 'Car'

- If fusion client is B and label is 'Plant', change label to 'Flower'

- If fusion client is C and label is 'Vehicle', change label to 'Car'

- If fusion client is C and label is 'Blossom', change label to 'Flower'

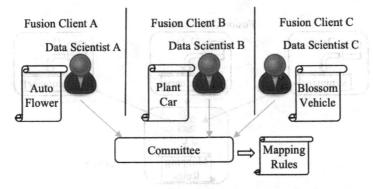

Figure 4.6: Reconciling value conflicts manually.

These rules can be defined centrally built upon a manual consensus, stored at the fusion server, and then be retrieved by each fusion client before the model training is done in order to reconcile all of the labels. If the data is featured, the different values of each column can be run through a similar process to get a set of consistent values for each item in the column.

4.2.2 *Summarization Approach to Reconciliation*

Since it may not always be viable to have an expert for each fusion client for a federated learning process, one may want to consider alternative approaches where there is only a single expert available and operational at the fusion server. This human expert can look at summarized results from each fusion client, and define the rules for translation for each fusion client. The process would begin by a software at each site which collects the set of all labels for every feature and sends them over to the fusion server. The human expert would then look at the aggregated set of labels across all fusion clients, and define the rules for relabeling that each fusion client ought to use. This process only requires one human expert, at the fusion server, who needs to explore the summarized data and define the appropriate translation policies.

Once the policies are defined, the fusion clients could retrieve the relabeling policies from the fusion server, and apply them to get a transformed view of the data that is to be used. Apart from the task of defining policies, the entire process can be automated. As a result, the approach shown in Figure 4.7 is more practical and effective than the approach shown in Figure 4.6.

The manual reconciliation process shown in Figure 4.7 will work well if there are a small number of well-defined labels for each of the different entries, or a set of well-defined values for each column of a featured data set. However, in practice, one may encounter situations where there are a large number of labels,

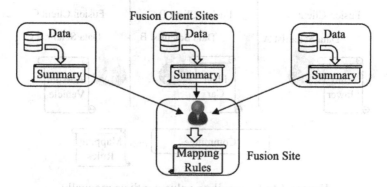

Figure 4.7: Summarization approach for label reconciliation.

and a manual reconciliation is not possible. Tools that can help the expert at the fusion server in the task of reconciling policies would be useful to assist in the summarization process. We describe some of these tools in the subsequent section.

4.2.3 *Cross-Site Confusion Matrix*

The cross-site confusion matrix can be a useful tool to identify the labels that are different at different fusion clients. This uses the concept of confusion matrix, which is frequently used in machine learning classification approaches to analyze and understand the performance of different algorithms.

The confusion matrix of an AI model shows the percentage of the prediction of the model that matches the true label of the data. Each row of the matrix represents the count of the predicted class for a data point while each column represents the true value of that data point. In some papers, the order of rows and columns is switched, but that does not make a material difference to the use of the confusion matrix. For the purpose of this chapter, we will consider the confusion matrix in which each entry in a cell of the confusion matrix represents the percentage of the actual instances in the test data which was predicted correctly by the model. In that case, the columns of the confusion matrix would sum up to be 1.0.

It is common practice in machine learning to divide available data into a training set and a testing set, train the model on the training set and use the confusion matrix to evaluate the resulting model on the testing set. Usually, when the training data used to train the model has the same labels as the testing data, the labels on the rows and columns of the confusion matrix would be the same.

If one were to train an AI model completely using the data available to fusion client A, one would get a model that converts any image into one of two classes, Auto or Flower. When it is applied on the data at the same client, A, e.g. by dividing it into a training and testing data set, it may result in a confusion matrix like the one shown in Table 4.8. Of the actual images in the class 'Auto', the model predicts 95% correctly and 5% incorrectly. Of the actual images in the class 'Flower', 97% are classified correctly and 3% incorrectly. In general, if training and testing data are not identical, it is rare for a model to be 100% accurate. However, if

Table 4.8: Confusion matrix example.

	Actual	
Predicted	Auto	Flower
Auto	95%	3%
Flower	5%	97%

Table 4.9: Cross-site confusion matrix for model trained at Client A and tested at Client B.

	Actual	
Predicted	Car	Plant
Auto	96%	2%
Flower	4%	98%

the accuracy is good enough to provide a business advantage, some mistakes are acceptable.

In order to check for label inconsistencies across the different fusion clients, one can use the confusion matrix, except by using it across multiple locations. Each of the fusion clients would train their own model purely using the local data, and send it to the other fusion clients. The fusion server may be used as a central point for the distribution of the models across to each fusion client. Upon the receipt of the model from another fusion client, each of the fusion clients would run the model on its data and create a confusion matrix. Note that, in this particular case, the model would predict labels that it was trained on, namely the labels present at the fusion client which provided the training data. However, the actual labels would be the labels present at the fusion client that is running the model. In this cross-site confusion matrix, the labels on the rows and columns of the confusion matrix would be different. The comparison of the confusion matrix would reveal how the different labels across fusion clients map to each other.

In the cross-site confusion matrix shown in Table 4.9, the model is trained using data at fusion client A, as shown in Figure 4.5 and tested on the data available at client B. As a result, the actual labels, shown as column headings in the table, are the labels used at fusion client B, namely Car and Plant. The row labels, however, are the values the model would predict, which is the labels used at fusion client A, namely Auto and Flower. The comparison of the entries in the two confusion matrix shows that the label of Auto corresponds to Car and the label of Flower corresponds to Plant.

A tool that computes the cross-site confusion matrix across all of the fusion clients can help the data scientist shown in Figure 4.7 with the task of identifying the right label mapping policies. For the specific example of three fusion clients that we are discussing, this approach can be used to identify the correspondence among labels at different fusion clients.

4.2.4 *Feature Space Analysis*

While the cross-site confusion matrix provides a good approach to determine labels mapping with each other when the labels at different fusion clients have a one-on-one correspondence, it will not work very effectively when the labels are overlapping with each other. It is possible that one fusion client may be using two labels for some type of data for and another fusion client may only be using one type of data. For the specific case of images, let us consider the situation where three fusion clients have images of lawn-mowers or automobiles that need to be differentiated between. Fusion client one may be using automobile images that are labeled with the name of the marker of the automobiles or the lawn-mowers, while a second fusion client may be using images that simply label them all as vehicles or lawn-mowers. Similarly, a third fusion client may be using labels that distinguish among the manufacturer names of lawn-mowers, but keeping all the automobiles together as one class. If a company manufactured both automobiles and lawn-mowers, the images at the first fusion client that are labeled by the name of the manufacturer need to be separated into either a lawn-mower or an automobile. Splitting these labels using the cross-site confusion matrix may not work well because a significant number of images in the label of the first one would map to both of the labels in the solution of the problem.

If the data is featured, examination of the feature space and overlap among labels in the feature space may provide a way to resolve the situation. The set of all the features that are taken together define a feature space, i.e. a multi-dimensional space which covers all the data points which provide the possible set of inputs. Each label covers some region in this feature space, defined by the different points that correspond to a specific output label.

To explain the approach, let us consider a data set shown in Figure 4.8. For ease of illustration, only two features are shown. The training data set would consist of points of format <value of feature 1, value of feature 2, output label>. In Figure 4.8, we are assuming that there are three labels marked as X, Y and Z. If we trace out the closest curve enclosing the points that correspond to each of the labels, they will trace out different regions in the feature space. A label may map to more than one region in the feature space, as shown by the label X in the feature space. These regions can be identified using various clustering algorithms [56].

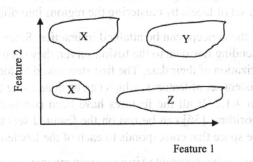

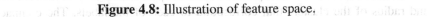

Figure 4.8: Illustration of feature space.

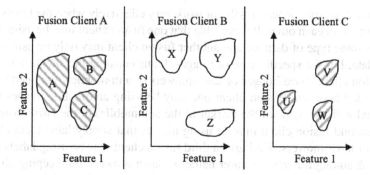

Figure 4.9: Feature space example with three fusion clients.

Assuming that the feature space is defined in a consistent manner across all the fusion clients, the overlap among the different regions at different fusion clients can be used to determine the correspondence between the labels. An example with three fusion clients with three labels each is shown in Figure 4.8. The three labels carve out regions that are marked with A, B, C in at fusion client A, as X, Y, Z at fusion client B, and as U, V, W at fusion client C. The regions at each fusion client are marked in different shades to highlight their difference.

If the regions carving out their labels at different fusion clients are super-imposed upon each other, we would get a diagram like Figure 4.10. The three regions overlap with each other in different manners. The regions of V, B and Y seem to overlap together fairly well, and can be considered one single label. The regions of C, W and Z also overlap with each other, and may be viewed as a single label. However, A overlaps with U and X, but U and X do not overlap. This means that perhaps A ought to be broken into two regions, one the same as W and the other the same as U.

Conceptually, if a summary sketch of each of the regions using a consistent set of features as dimensions is sent to the fusion server, then the fusion server can examine the overlap among the regions of each fusion client, and determine how to define a canonical set of labels by clustering the regions into different groups.

Let us examine how the concept can be attained in practice. Since fusion clients are not going to be sending raw data to the fusion server, they need to agree upon a consistent summarization of their data. The first step would be for all the fusion clients to use a consistent set of features, which can be done using the techniques described in Section 4.1.2. If all the features have been consistently scaled, a set of clustering algorithms [56] can be run on the featured sets to identify the regions in the feature space that corresponds to each of the labeled classes.

The identified clusters can be represented in a compact manner, e.g. as the centroid and radius of the clusters that correspond to each of the labels. The compact

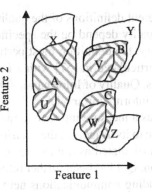

Figure 4.10: Superposition of feature space of three fusion clients.

summary of data can be sent over to the fusion server, which can examine the overlap among the different labels. This overlap can be done by running another round of clustering algorithms which can be run at the fusion server with the following steps (i) pick the least radius among all the reported clusters (ii) represent each cluster with the larger radius as multiple clusters, each cluster being represented by a centroid with the selected radius size picked to cover the entire original cluster and (iii) run a clustering algorithm among all the centroids to find the groups that overlap.

The overall process would be identical to the approach shown in Figures 4.9 and 4.10. Depending on the clustering algorithm that is chosen, the shapes generated for each region may have a regular shape instead of the irregular ones used in the illustration.

Some clustering algorithms work better if another step, such as principal component analysis [13], is done prior to the clustering process. If such an algorithm is used, one should use a consistent mechanism for the conversion to principal components across all of the fusion clients. In these cases, the use of a naive federated learning algorithm, such as function fusion across all of the fusion clients, 3.3 can be used to come up with a consistent principal component analysis approach across all of the clients.

4.3   Eliminating Poor Quality and Low Value Data

One of the challenges in using data from multiple fusion clients is that the data collection processes and data management procedures at all fusion clients need not be consistent. This results in data that has a variable quality across the clients. Incorporating data with poor quality can result in bad AI model. Therefore, one of the pre-processing tasks would be to understand clients that have poor quality data, and eliminate instances of the same.

While there are many different definitions of the quality of data that is available, and many definitions of quality depend on the specific type of data, e.g. on image data or satellite imagery, the definition most pertinent for machine learning in business contexts was articulated in [57] with a focus on sensor information fusion. It defined two terms, Quality of Information (QoI) and Value of Information (VoI), along with an information model to describe the attributes required for both. QoI was defined as a measure intrinsic to the information which can be measured without reference to its usage, i.e. the decision making task that utilized the information. VoI was articulated as a measure of merit which depended on the usage of the information. QoI has been further refined and applied for various application domains, including communications networks [58], wireless sensor networks [59], and mobile crowd sensing [60]. VoI has been further defined and applied for Internet of Things [61], data retrieval [62] and machine learning for coalition operations [63].

Quality is an intrinsic property of the data that is being collected, and it can be estimated by examining the properties of the data independent of the business process within which it is used. The value of the data depends on the business process, namely how that data would be used. The value is also incremental, i.e. the value of a piece of data is dependent on what additional data is already available to the model or the business process. If a new data set is a duplication of the data that is already available, it may not add much value despite being of very high quality.

For defining quality and value of data, the operations in the Learn→Infer→Act cycle can be modeled as the process shown in Figure 4.11. There is some function pertinent to a business process which needs to use an AI model in order to perform or improve its function. The ground truth is the function in the real-world which the training process is trying to model. In order to learn this function, training data is collected. The data that is collected encodes within itself a representation of the ground truth, however, mistakes may be made during the collection process.

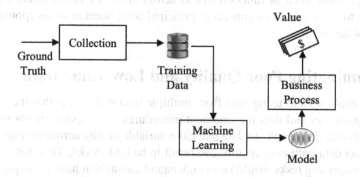

Figure 4.11: Abstract process of machine learning for quality definition.

As a result, the training data may encode a function which is different from the ground truth. The model in itself is an alternative representation of the function. When that model is used within the business process, some value is generated. The value of the training data or the model is the value that is generated by the business process. The value of the business process may be the additional revenue that is collected, cost that is saved, or any other associated metric which needs to be monitored and improved upon.

The quality of data or the quality of model is its ability to capture the ground truth correctly. The quality of the training data can be measured by how close it is to the ground truth. The quality of the model can also be estimated by how close it is to the ground truth. The ground truth, the training data and the model are all different representations of the same function, and in principle can be compared to estimate how different they are. The smaller the difference, the better the quality of data. The challenge in most practical business applications is that the ground truth may not known. As a result, an absolute measure of quality of data is not feasible, and proxy measures have to be deployed.

The value of information is the contribution made to the business process value function by the collected data. This value could depend on several factors beyond the model, as well as the difference the model provides over and beyond the value which was not already available.

A detailed discussion of different approaches for estimating quality and value of information can be found in references such as [64, 63, 65, 66]. Without going into more technical aspects of measuring quality and value, we would assume that good approaches to measure quality and value of information exist. We would instead focus on how the metrics associated with them can be used to improve the data available for federated learning.

4.3.1 *Reputation-based Data Selection*

One approach to improve the quality of data is to assign quality and/or value metrics at the granularity of the fusion client. In that case, one can determine whether or not to include data provided by any of the fusion clients based on how trustworthy that fusion client is. In some cases, there is a lack of trust among the different parties or some of the fusion clients may be considered more trustworthy than others. For example, in a coalition among different countries, some of the countries with a separate military alliance between themselves may trust each other more than those of members not in the alliance to maintain good quality data. A specific instance would be when NATO countries conduct a peace-keeping mission in Africa, the coalition could consist of NATO member countries and African countries, but the NATO countries may not consider data from the African alliance members to be equally reliable. Another instance would be in sharing health and disease-related information among different nations for a task such as combating a pandemic, where almost all the countries in the world would be cooperating together, but rich countries with more advanced health-care infrastructure may consider data from countries with less resources and poorer health-care infrastructure as being less reliable.

If there is an approach to assign a reputation to different fusion clients providing data for the federated learning process, data from fusion clients with poor reputation can be dropped from the process of federation. The fusion server may or may not drop the fusion clients explicitly, choosing to let such clients participate but ignore the models provided by the low reputation clients during the model fusion process. The silent approach gives the low reputation clients a model that is created by combining the models from the high reputation clients. There are many techniques for assigning reputation to different clients in a distributed computing contexts [67], and any suitable reputation computing system can be used.

A reputation-based mechanism for ignoring data or models from selected clients would work well in environments where there are a large number of sites with data, and one can create comparison metrics for assigning reputations. This could work in settings like consumer federated learning (See 2.3.1) where there would be thousands or millions of clients with data. However, in the context of enterprise federated learning (See 2.3.2), there may not be sufficient number of locations to afford dropping all the data from one of the clients.

Because of the large granularity required for per-site selection, reputation-based data selection may not be a viable approach for Enterprise Federated Learning, which is the primary scope of this book.

4.3.2 *Value-based Data Selection*

In value-based data selection, data for federated learning is selected from fusion clients based on the value that each additional data point is contributing to the task of model building. The value judgment can be made based on the contribution that is being made by each of the clients in the task of federated learning. When the main task is focused on training an AI model, one good estimate of the value of the data set is the amount of new data that is contributed by each of the fusion clients [68].

In an analysis of federated learning for coalition operations [69], a theoretical analysis of the value of data obtained from different coalition partners is obtained under some simplifying conditions. The analysis is based on exploring whether there is value in obtaining new training data from a partner based on the currently available data for training the model. Each partner has some data that is noisy and some data that is good and noise-free. The analysis shows that it is better to get additional data from the partner as long as the partner is providing some new data that is free of noise over and beyond the data that is already available. This is based on the assumption that for each partner, the noisy data is small proportion of the overall data. Thus, if there is some data which is not already present locally, it is better to get the information from another partner.

The situation is shown graphically in Figure 4.12. It is assumed that there is some local data at one site, and the site has the choice of accepting data from the partner site. The shaded regions show the part of data that is noisy, either locally or at the partner site. While the noisy data would be a relatively minor fraction, the figure exaggerates it in relation to the good data that is not shaded. Part of the good data available at the partner and part of the noisy data available at the partner is common with the data is present locally. However, there is some part of the data that is good and not present locally. If a substantial amount of good data is obtained from the partner, it would be worthwhile to obtain that data from the partner.

The challenge in real-life is the task of determining how much of the data is common and how much of the data from a partner is new. One needs to make this determination without exchanging data with each other. The basic idea to assess the similarity between the data present at different sites would be similar to that described for feature space analysis for resolving value conflicts described in Section 4.2.4. The labels can be ignored for the purpose of determining the value of new data, and the measure would be the additional area of feature space that is covered by data belonging to the new partner.

The assessment of the value contributed by partners can be measured on the basis of the entire data available from the partner, or on the basis of a selected subset of the data provided by the partner. The selection of the subset can be done on the basis of labels, e.g. one can collect data belonging to a specific label from a partner only if the partner has some new data that is not already provided by the other parties, essentially accepting data only if the number of points belonging

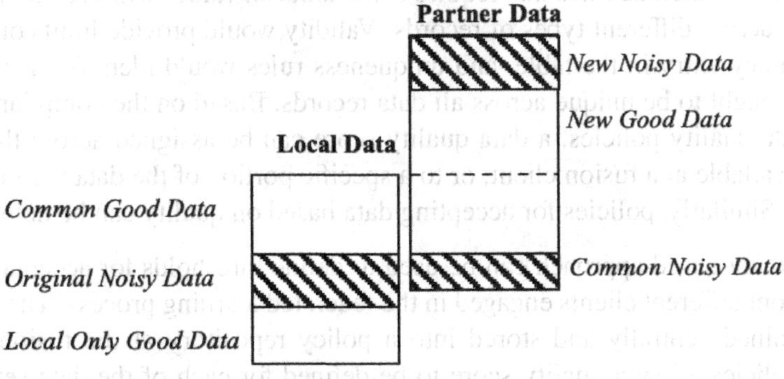

Figure 4.12: Value-based data selection from partners.

to the label increases. In this case, some labels from a partner are accepted while other labels from the same partner can be rejected.

The assessment of value can also be done on random partitions of data to select those data points which do not have a substantial overlap. In this case, before the training process begins, each of the clients would divide their data into partitions of a predetermined size, and the value offered by the data partition compared to data available from other clients would be provided. This would eliminate data partitions that do not provide good value or are already available from other partners.

4.3.3   *Policy-based Quality Improvement*

A policy-based quality assessment method can be used to select portions of data from selected fusion clients that do not match quality requirements. Policies that define different metrics for data quality can be defined at the fusion server. A policy-based data quality architecture citegrueneberg2019policy has several advantages over other means of defining data quality. It allows a flexible definition that can be easily customized for any specific AI model building task, any specific business process application, and can even provide automated means to customize the policy for different fusion clients.

Data quality policies assign a quality score to any piece of data based on a variety of rules. These rules are structured to obtain quality scores on different categories, such as completeness of data, consistency of data, validity of data, or uniqueness across data. The final quality score is obtained by combining the scores along each individual category.

An intuitive feel for various categories can be obtained by mapping these attributes to data stored in a tabular representation, such as a spreadsheet or a relational database. Each row in the table is a record or data point. Completeness would indicate that values are not missing from specific columns in the record. Consistency would include rules that require constraints on values within a record, or on values across different types of records. Validity would provide limits on entries within any entry in the table, and uniqueness rules would identify those values which ought to be unique across all data records. Based on the compliance with the data quality policies, a data quality score can be assigned across the entire data available at a fusion client, or to a specific portion of the data records at the client. Similarly, policies for accepting data based on quality can be defined.

The policy-based approach can be used to define thresholds for acceptable data sets from different clients engaged in the federated learning process. The policies are defined centrally and stored into a policy repository at the fusion server. The policies allow a quality score to be defined for each of the data sets. Each fusion client would retrieve the policies relevant for their needs from the policy

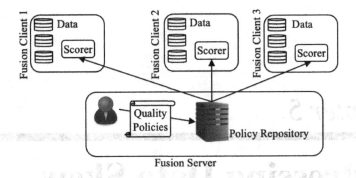

Figure 4.13: Policy-based quality improvement.

repository, and use them to calculate the quality score for each data set. The data sets passing the thresholds for acceptable quality will be the only ones each client would use to train its model.

The architecture for this policy-based approach is shown in Figure 4.13. A policy-based data scorer looks at various partitions of the local data and assigns them a quality score based on policies retrieved from a central location. The scorer can decide the local subset of data that ought to be used in the task of federated learning.

4.4 Summary

In this chapter, we have examined the approaches that can handle the problems of data mismatch that arise in the context of learning across data sets that are distributed across different fusion clients. We have identified the different types of mismatches that could exist across different fusion clients. These include differences in formats, differences in how the same concepts are named across different fusion clients, and differences in the quality and value of data that is available across different fusion clients. We have explored how one can address these mismatches by using constructs such as conversion graphs, cross-site confusion matrices, a centralized definition of rules for transformation of data, and the enforcement of a consistent set of quality policies.

The application of federated learning in any realistic context with differences in the data would require the use of these constructs. Some of these constructs themselves depend upon running auxiliary machine learning algorithms, like clustering as a sub-component. The creation of any solution would require many of the basic components to be arranged into an end-to-end work-flow that is customized for the business problem, and the right set of data management algorithms to be selected and applied before the actual task of federate learning begins.

Chapter 5

Addressing Data Skew Issues in Federated Learning

Data skew refers to the situation where the training data available for learning at different fusion clients has very different characteristics. In this chapter, we look at approaches to address issues caused by such a skew. For the purposes of this chapter, we assume that the data consistency and data normalization procedures described in Chapter 4 have been applied. Specifically, issues dealing with data formats, consistent normalization across fusion clients, and managing data quality have been dealt with. As a result of these steps, each fusion client would have data in the same format, normalized in a consistent manner. Data at all the fusion clients has comparable quality that is acceptable to all. We will also assume that the fusion clients have an implicit trust in each other and the data server site. This assumption is realistic for most enterprise settings. Nevertheless, in Chapter 6, we consider some of the approaches to handle the situation when this trust may be limited.

If we are lucky and the data is more or less randomly distributed across all of the fusion clients, one of the approaches outlined in Chapter 3 would work well. In most real environments, data distribution across fusion clients will be skewed. Therefore, we must have approaches to make the model perform well even when the data distribution is skewed.

One type of data skew may happen because different fusion clients may be collecting data collected under different underlying assumptions. Another type of data skew can happen because the data collected at some clients may be missing some of the required classes. One case which is particularly challenging for machine learning is when the data is partitioned completely among different fusion clients.

To explain partitioning within featured data, let us consider the structure of this data stored in a tabular format. Such data consists of several rows and columns. In problems like classification or mapping, there is one special column (the output column) which contains the class or label of the entry.

Let us assume that the centralized version of the data would have the contents as shown in Table 5.1. In this table, 6 records are shown for illustrative purposes, where the data belongs to three output classes with labels L_0, L_1 and L_3. The data consists of three features F_1, F_2 and F_3, and the index shows a distinct number for each record.

The centralized data would be obtained if all the data from all fusion clients were collected at a central location, the redundant data records eliminated and a single index created to identify each record uniquely. While such centralization is not feasible in many real-life environments, which is the primary reason for federated AI, the thought experiment of collecting the data would be useful for illustrative purposes.

When the data is present at different fusion clients, there is likely to be redundancy of data among different fusion clients, and it is unlikely that there is a consistent index across all of the fusion clients. We will however, use the index to illustrate a hypothetical unique identity of each record. The data at each of the different fusion clients would be a subset of the centralized data that is shown in Table 5.1.

One possible split of the data among three different fusion clients may be as shown in Table 5.2. This is a relatively nice partitioning of the data. Each of the fusion clients has data that belongs to each of the three labeled output classes. Each fusion client has all the features that are present. The index entry will not be present at any fusion client in a real setting, or at least it will be independent. However, it is put there to have an easy correlation with the centralized data in Table 5.1.

Table 5.1: Example for partitioned data.

Index	F_1	F_2	F_3	Output
1	A	X	0.3	L_1
2	B	Y	0.2	L_2
3	C	Z	0.3	L_2
4	A	X	0.4	L_3
5	B	X	0.8	L_1
6	B	Y	0.9	L_3

Table 5.2: Nice partitioning for example for partitioned data.

Fusion Client A				
Index	F_1	F_2	F_3	Output
1	A	X	0.3	L_1
2	B	Y	0.2	L_2
4	A	X	0.4	L_3

Fusion Client B				
Index	F_1	F_2	F_3	Output
3	C	Z	0.3	L_2
4	A	X	0.4	L_3
5	B	X	0.8	L_1

Fusion Client C				
Index	F_1	F_2	F_3	Output
1	A	X	0.3	L_1
2	B	Y	0.2	L_2
6	B	Y	0.9	L_3

Table 5.3: Data partitioning example.

Fusion Client A				
Index	F_1	F_2	F_3	Output
1	A	X	0.3	L_1
5	B	X	0.8	L_1

Fusion Client B				
Index	F_1	F_2	F_3	Output
2	B	Y	0.2	L_2
3	C	Z	0.3	L_2

Fusion Client C				
Index	F_1	F_2	F_3	Output
4	A	X	0.4	L_3
6	B	Y	0.9	L_3

If the data set is partitioned as shown in Table 5.2, the federated learning algorithms described in Chapter 3 will work reasonably well. While the data is divided into multiple fusion clients, data at each fusion client has some overlap pairwise with data present at all of the other clients, each output class is present at each fusion client, and they are likely learning the same function.

In real-life, it is common to encounter scenarios where the division of data across fusion clients is not always designed to play nicely with the algorithms used for federated learning. One specific type of situation that arises is when the data is divided into different fusion clients so that the data at each fusion client is missing some of the classes. An extreme form of this is shown in Table 5.3. In this particular distribution of data among the clients, each of the fusion clients has data belonging to only one class. When data is partitioned in this manner, naive algorithms for federation fail to perform well [70]. The partitioning need not be this extreme to impact performance, there are other partitioning patterns of data that can pose similar challenges for federation, e.g. some classes are present at only one fusion client, or every fusion client may have multiple classes but some fusion clients are missing most of the classes.

A similar partitioning of data can also occur in featureless data. There could be fusion clients which are missing data belonging to some of the output labels.

5.1 Impact of Partitioned and Unbalanced Data

In this section, we would examine the impact of partitioned or unbalanced data in a few situations, and discuss the underlying causes why partitioning of the data causes a problem in the task of federated learning.

5.1.1 *Data Skew Issues in Function Estimation*

One of the ways to understand data skew in distributed data is to explore the problem from the perspective of function estimation. Real-world functions are complex, and different fusion clients may be collecting data points that operate under different conditions under which the function is measured. When different fusion clients are learning a different function, naive algorithms would not be able to compose or represent those functions properly.

An example where different fusion clients may end up learning different functions is illustrated in Figure 5.1. The figure has four diagrams, and shows a situation that may arise in the real world when data is collected independently at two different fusion clients. All the data collected at the two fusion clients along with the ground truth is shown in the diagram on the top left. The ground truth is the true relationship that should hold between the input (x) and output (y) and is shown with dashed lines. For ease of illustration, only one input is shown, but the discussion can easily be seen to generalize to more complex inputs with several features. The stars mark the data points that are available at one of the fusion clients (client A) and the circles mark the data points that are available at one other fusion client (client B). When either of the two fusion clients estimates the

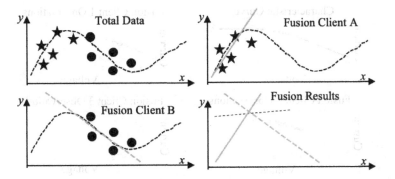

Figure 5.1: Different views of the data.

function to match their data, the function learned would be based on the data set available at that fusion client.

On its own local data, fusion client A would learn a linearly increasing function between x and y, as shown with a thick grey line in the top right diagram. It is a natural expectation that the function estimation allow the system to extrapolate relationship between the input and output. Similarly, fusion client B, when estimating a function using only its local data, would learn a linear decreasing function, like the one shown with a thick dashed grey line in the bottom left figure. If federated averaging techniques are used on this data, the net result would be a linear function that is shown in the black dashed line on the bottom right figure. The summation of the two functions will not look anything like the ground truth.

The problem arises from the fact that both fusion clients collected data under different conditions. As a result, each fusion client has data that represents a different function. The function at each of the fusion clients is a valid relationship between the input and the output, however each relationship holds true under a different set of conditions. For the example shown, the condition is the range of the input value. A simple averaging of the learned functions (i.e. the model), assuming that they all represent the same relationship, may not always be the right approach to take.

Let us consider an example in distributed modeling, i.e. the behavior of an electronic component [71]. The situation is shown in Figure 5.2. There are four sub-images in the figure, the left top showing the characteristic curve of an electronic component which is the ground truth, and the other three showing data points that are collected by different fusion clients testing the component. The experiments at each of the fusion clients has focused on one of the regions of the operation of the electronic component. As a result, the characteristic curve predicted by each of the fusion clients would be very different.

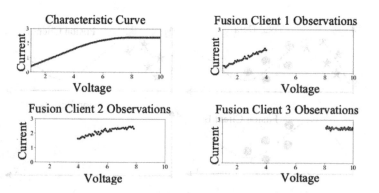

Figure 5.2: Data skew illustration for function estimation.

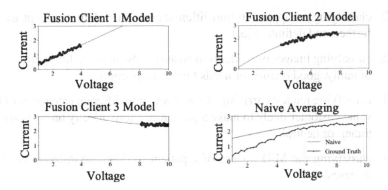

Figure 5.3: Impact of data skew on function estimation.

The result of naive federated learning algorithms described in Chapter 3 on each of these models can be off-target because the data at different locations is related to different regions and different characteristics. The rationale would be similar to that illustrated in Figure 5.1.

The result of fitting a function on the three different data sets and using naive federated learning on them is shown in Figure 5.3. The figure shows four sub-diagrams. Fusion client 1 has collected data for the lower end of the characteristics curve, and it estimates a relationship between the input (Current) and the output (Voltage) which is a linear increasing relationship. Fusion client 2 has collected data from the middle of the operating range of the component and it has learned a relationship which is closest to the ground truth among each of the three clients. Fusion client 3 has learned a relationship that shows a decreasing relationship between the current and the voltage. With the possible exception of client 2, no client has a reasonable estimate of the ground truth. When the results from all the clients are averaged together, one gets a result which does not look like the ground truth.

In order to combine the estimated functions properly, we need to incorporate awareness that the different fusion clients are learning functions applicable for different ranges of the input into the task of combining models.

5.1.2 Label Partitioning Issues in Classification

Classification is a very common task in AI-enabled functions, which primarily deals with the task of mapping any input into one or more classes. There are many instances of useful business processes based on classification, which include the following, and many more:

1. classifying credit card applicants into different levels of risks, e.g. 5 classes of definite reject, high risk, medium risk, low risk, and definite accept

2. classifying job applicants into different categories, definite accept, examine further, and definite reject

3. classifying images of products in an assembly line into different categories of faulty, good quality, or needs further inspection

4. classifying requests arriving at a web-site into different categories of high-value customer likely to make a purchase, a low-priority browse-only customer, or neutral

5. classifying the MRI images of a patient to diagnose different classes of diseases.

Given the wide applicability of classification in many business use-cases, several algorithms for classification that work well when the data is centralized have been developed. Classification can be done in either a supervised or an unsupervised manner. In supervised classification, several instances of training data set with labels are available, and the AI models learn how to differentiate among the different classes based on the training data. In unsupervised classification, the data is not labeled, but clusters of similar images can be defined and labeled. The additional challenge with skewed data applies to both supervised and unsupervised classification, but would be easier to explain in terms of supervised classification problems.

Normally, if the task requires data to be classified into some N number of classes, and each fusion client has some instances of each of classes, the naive federated learning algorithms described in Chapter 3 work well. However, when the classes are missing from a set of fusion clients, the naive federated learning algorithms are less adept. A specific situation is illustrated in Figure 5.4, which shows three fusion clients with their data and labels. The first fusion client has data that belongs to two classes, A and B, the second fusion client has data that belongs to two other classes, C and D, and the third fusion client has data that belongs to yet

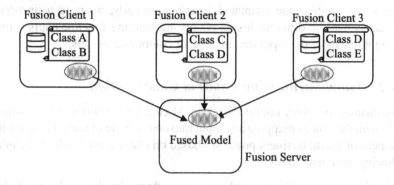

Figure 5.4: Illustration of data skew in classification.

another two classes, D and E. Each of the fusion clients is training an AI model on its own data, and then sending it over to the fusion server for combination as per the configuration in Figure 3.1.

The problem in a situation like the one shown in Figure 5.4 is that each fusion client trains the models so that any data is mapped into one of the labels that are available in the local data. AI models for classification learn to map any input to one of the classes they have been trained upon. They usually lack the ability to detect the fact that a data input may belong to a new class.

In this particular case, the model trained at fusion client 1 would map any input to labels A and B, fusion client 2 would train its model so that any input is classified into labels C and D, and fusion client 3 would classify any input into the labels D and E. The only class which has data present at more than one fusion client is class D. Apart from inputs which are mapped to class D, no two models will predict the same class for any input. With a high probability, one would expect any input belonging to class A to be classified as A by model trained by fusion client 1, but it would be classified into any of the classes C or D by the model at fusion client 2, and it would be classified into D or E by model trained at fusion client 3. The three models have learned a very different function, and averaging them in a meaningful manner would be extremely difficult.

This behavior of models predicting only what they have been trained upon is common across many types of AI models, including neural networks. Even when the architecture of the neural network is the same, i.e. they are using the same number of levels of neurons, and the same number of neurons at each level interconnected in an identical network, the naive averaging of parameters across different fusion clients in a scenario like the one shown in Figure 5.4 is not likely to lead to a good model. The reason for this is that each of the fusion clients is training its neural network to adjust weights to learn a different class of function, one that maps input into different classes. Averaging the weights across different functions results in something weird because there is no fundamental reason for averaging these weights to result in something meaningful. The fundamental rationale for averaging across neural network weights and having it work is that the system is learning the same function at all fusion clients, and different views of the different functions can be averaged together to end up with a meaningful result.

A discussion of the impact of missing classes across different fusion clients for federated learning among different federal agencies [70] can be useful to illustrate the challenges in federated learning when classes are partitioned. The specific situation considered was that of the classification problem where data was divided into ten unique classes, which was divided into ten different fusion clients (federal agencies in the case of the original paper). When each of the fusion clients had some data belonging to each of the ten classes, the naive federated learning

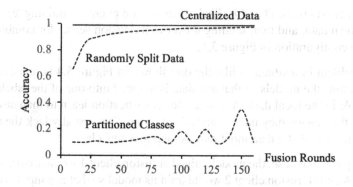

Figure 5.5: Impact of data skew on classification.

algorithms performed very well. However, when data was split so that each class was present only at one of the fusion clients, the performance of the resulting model was very poor.

The challenges posed by partitioned classes is shown in Figure 5.5. The effectiveness of classification into ten classes is measured using three different scenarios. In each scenario, the effectiveness of a model is measured by creating a model over a training data set, and evaluating it over a different testing data set. The training data set is identical across the scenarios. In the first scenario, all data is maintained at a centralized location, i.e. data from each of the ten participating fusion clients is collected at a central location and a model trained on that. In the second scenario, the training data is not moved to a central location, but it is assumed that each fusion client has a random partition of the training data, i.e. each fusion client has some instances belonging to each of the ten classes. Federated averaging algorithm, an instance of the naive federated learning algorithms discussed in Chapter 3, is used and federation is done over several rounds of averaging over the different instances of the model. In the third scenario, the same algorithm is used, but the training data is distributed so that each of the ten fusion clients has data belonging only to a single class.

The effect of this partition on naive federated learning algorithm performance is disastrous. The combined model is not able to perform well despite several rounds of federated averaging. In effect, the approach of combining models using federated averaging does not work.

There will be partitioned data in almost every real-life business environment. Therefore, it is important to explore approaches that can handle the impact of the partitioning of different data. In the next two sections of this chapter, we look at some of these approaches.

5.2   Limited Data Exchange

The motivation for not sharing data and using federated learning varies depending on the context of the application, and in some cases any data exchange may not be permitted due to security or privacy concerns. In other cases, data exchange is not disallowed, but it is relatively expensive or time-consuming. In the latter case, it would be okay to exchange some amount of data among different fusion sites, as long as it is within a threshold permitted by cost or transfer time considerations.

In those cases where some data exchange is permitted, there may be a simple yet effective mechanism to deal with the challenge of data skew. If the challenges are arising due to insufficient classes being present, or due to data belonging to different ranges of the input space (as in function estimation), one may be able to exchange some data and allow each of the fusion clients to have some instances of the data that is present.

For the case of function estimation, each of the fusion clients can determine the input range for which their data was collected. If the input range is different, each of the fusion clients can exchange a limited amount of data so that they all have a similar input range. This limited data exchange needs to be done as a pre-processing step before the actual process of federated learning happens. Limited data exchange can result in the improvement in the task of federated learning by increasing the applicable ranges where the function is applicable.

To perform the limited data exchange, each fusion client can send the ranges of the input over which it has collected data to the fusion server. One way to send this range would be as the upper and lower values on all the input features. Other statistical measures on the input features, e.g. the 5th percentile and the 95th percentile, can also be used. The fusion server uses the ranges to determine if any fusion client is missing data in some range of the input space. If all the fusion clients have collected data over the same range or very similar range of input values, no further action is needed. However, if any fusion client has gaps in its range of data collection, the fusion server can identify those gaps. It can ask other fusion clients to provide some data samples in that gap to their peer with the gap. This ensures that all fusion clients have the same range of collected data, are computing their function estimates over the same range, and each fusion client ends up computing different approximations to the same function.

We can illustrate that adjustment in the range by exploring what the exchange of data would allow in the hypothetical situation illustrated in Figure 5.6. This is a repetition of the hypothetical example shown in Figure 5.2. However, as part of the limited data exchange, some of the points are exchanged among the different fusion clients. The exchanged points are shown as grey in the data for fusion clients A and B, respectively. With this exchange, the range over which the function is learned changes for both fusion clients. The original range is shown

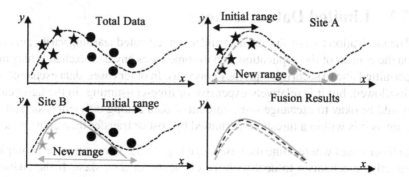

Figure 5.6: Impact of data exchange on function estimation.

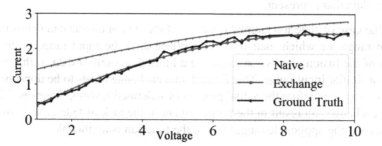

Figure 5.7: Impact of data exchange on electronic component model.

in black and the extended range is shown in grey arrows in the data figure for the two fusion clients. Since the range is now approximately the same for both of the fusion clients, they learn different approximations of the same function, and the federated averaging process gets them to a function that is a much closer approximation to the ground truth.

We can also examine the impact of this limited data exchange in the example of the electronic component model discussed earlier in this chapter (refer to Figure 5.3 in Section 5.1.1). In a repeat of the experiment with the three different fusion clients with three different perspectives on the component data, each fusion client exchanges information and thereby expands the range of functions for which data is available at each fusion client. The net result of this exchange is a greatly improved estimation of the function, as shown in Figure 5.7.

The limited data exchange works well for the case of partitioned data for classification as well. In these cases, the fusion sites can send statistics about the number of data points they have corresponding to each label/class to the fusion server. The fusion server can determine if any fusion client is missing some classes, and ask other fusion clients having data for those classes to provide some. This ensures

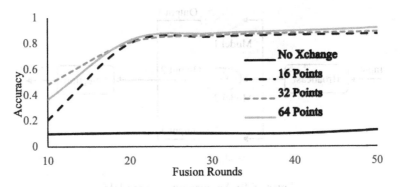

Figure 5.8: Impact of data exchange on classification.

that each fusion client is training models that predict all of the classes as output, and the model fusion process can be applied. Alternatively, each fusion client can report some of the data points it has on a per class basis to the fusion server, who then consolidates all of the points and sends them over to every fusion client. On the receipt of this data, each fusion client can add to its local data for training models.

The impact of data exchange on the effectiveness of classification models with partitioned data is shown in Figure 5.8. The dark black line shows the impact without any exchange, which is the same as the result shown in Figure 5.5. The performance of the AI model trained with partitioned data is not very good. The other three lines show the result with the exchange of a few instances of each type of class. There are only a small number of data points that are exchanged, in comparison to the data points that are present. Each fusion client had about 6,000 instances of each class, but the amount of data exchanged is 16, 32 or 64 instances per class. Even with a minor exchange of the classes, the accuracy metric of the model increases significantly with very few rounds of federated averaging fusion.

The results from the limited data exchange provide an approach to improve the quality of federated learning models in situations when the data is partitioned. In situations where real data can not be exchanged, providing a few synthetic samples of the missing labels would have a similar impact on the quality of the federated model that is created.

5.3 Policy-based Ensembles

Another alternative to handle the data partitioning problem is the use of an ensemble. An ensemble is a way to use multiple models to solve the same problem. The assumption in an ensemble is that all the models are trained to do the same

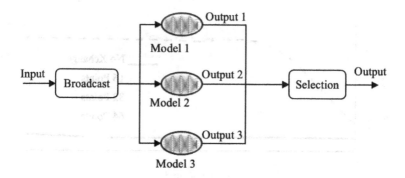

Figure 5.9: An ensemble of models.

task, i.e. they use exactly the same input and produce the exact same output. Furthermore, the ensembles are assumed to be independent.

In the ensemble approach [48], the same input is passed through multiple models, i.e. the input is broadcast to all of the models, and the output produced by each of the models is examined and one of the possible outputs selected, e.g. by majority voting to determine the final outcome. The ensemble approach can work better than any individual model, and is better able to deal with the variations in the input, assuming that each of the models in the ensemble is independent of the others. An ensemble with three models is shown in Figure 5.9, along with the components to broadcast the input and combine the models output.

Ensembles provide an alternative way to combine models from different fusion clients involved in a federated learning session. Instead of trying to combine models that are provided by each of the different fusion clients, the fusion server creates an ensemble of different models. One advantage of the ensembles is that the individual models need not conform to the same architecture. The ensemble approach can be used to combine output from many different types of models, which may have varying architectures and varying internal model structures.

Since the identity of each of the models is maintained separately, meta-data concerning the training of each model can be maintained for each of the models making up the ensemble. This information can then be used to judiciously combine different predictions made by individual models that make up the ensemble. One such approach would be the use of policies.

A policy [72] is a collection of rules, each rule defined by a condition-action pair. Each rule defines the action to be taken when the set of conditions specified in the rule are satisfied. The policy can be defined directly as a set of such rules, or defined as a higher set of constructs which are then refined into the set of rules. By defining policies to take different actions, the operation of any system can be modified to be different under different conditions. Policy-based management has

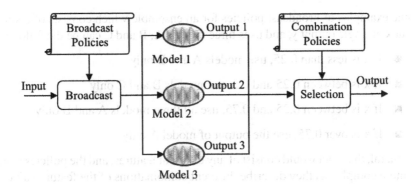

Figure 5.10: A policy-based ensemble.

been used to simplify the management of systems and networks in many different environments [73], ranging from management of networks to management of caching systems and applications. It stands to reason that policies can be used to control the operation of ensembles as well.

A policy-based ensemble operates like an ensemble but with two additional components, both of which are controlled by policies, as shown in Figure 5.10. The first component deals with the selection of the models within the ensemble. Under the control of policies, one need not use all of the ensembles in the policies, but a subset can be selected based on a set of conditions. Under some conditions, all of the models in the ensemble can be used. Under other conditions, a smaller subset of models can be used in the ensemble. Under yet other set of conditions, only one model will be used. The conditions selecting the subset of ensembles to be used are defined by means of the broadcast policies shown in Figure 5.10.

The second component controlled by the policies deals with the combination of the output from each of the selected model within the ensemble. Instead of just using the majority weight, or a weighted average, the combination of the outputs can be done more flexibly using a set of combination policies. Suppose the combination is done by assigning a set of weights to the output of different models. One use of combination policies may be to change the relative weights of the models based on a set of conditions.

Both combination and broadcast policies are defined using a set of conditions and corresponding actions. For the broadcast policies, the action is simply the selection of the different models available in the ensemble. In other words, if the ensemble consists of N models, the selection action can be viewed as a vector containing N Boolean entries, each indicating whether or not the corresponding model is selected. The conditions for broadcast policies can be defined either using the values of the input to the models, the characteristics of the models that are to be used, or information about the training data used for the model.

Some examples of broadcast policies for an ensemble which consists of a single input x, a single output y, and uses three models A, B and C can be the following:

- If x is less than 0.25, use models A and B only
- If x is between 0.25 and 0.5, use models B and C only
- If x is between 0.25 and 0.75, use all three models A and B only
- If x is over 0.75, use the output of model A only

In general, the input would consist of any different features, and the policies would be more complex as they describe different combinations of the feature values.

Each combination policy would have a condition and an action. The condition would be a logical expression defined over the value of the input to the ensemble, the type of model, the output of any model, or any attribute of the meta-data about the model. The action would be the weights assigned to the output of each model in the ensemble. For an ensemble which takes in a single input x, predicts a single output y, and is composed of three models A, B and C, combination policies may look like the following:

- If any model has predicted y to be less than 0.25, only use weighted average of models which have predicted y to be less than 0.5
- If x is between 0.25 and 0.5, use the weights of 0.25, 0.5 and 0.25 to combine models A, B and C, respectively
- If x is between 0.25 and 0.75, use models A and B output only
- If x is less than 0.25 or over 0.75, use the weights of 0.5, 0.25 and 0.25 to combine models A, B and C, respectively

A more detailed discussion of policy-based ensembles and their usage can be found in [74] and [75]. They have been shown to be an improvement on other approaches to create models for a variety of AI models.

Let us examine the impact of policy-based ensembles on the performance of the function estimation model for electronic components described earlier. Each model is characterized by the upper and lower bounds of the input (Voltage) of training data that was used to create the model. The broadcast policy in this case could state that only the models that were trained on data covering the value of input ought to be used within the ensemble. In case there is no such model which is used, two models which are closest in their range to the input value should be used and their output averaged. This provides an alternate mechanism for the output models from three fusion clients to be combined.

The result of combining the three models with data shown in Figure 5.2 and Figure 5.3 is shown in Figure 5.11. When the result of a policy-based ensemble

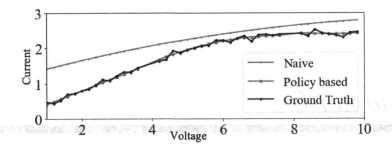

Figure 5.11: Impact of policy-based ensembles for function estimation.

is compared to the naive approach of combining output, one can see that the policy-based ensemble results in a more accurate model. The result is not truly surprising, since policies allow preference to be given to models that are better suited under different conditions.

Policies for combining models in an ensemble need not be defined manually. They can be generated automatically by examining the properties of the data that was used to train the models [74, 75].

Policy-based ensembles also provide a good solution to some of the data synchronization issues that are discussed in Chapter 7.

5.4 Summary

In this chapter, we looked at the challenges that arise when data is skewed. When data is skewed, different fusion clients may be learning different functions which should not be averaged naively. Data skew can result from situations where different fusion clients are observing the information under different assumptions or operating environments. It can also result when some output classes are missing from some of the fusion clients.

One approach that works well for addressing data skew is limited data exchange if permitted within the business context. Data exchange expands the scope of function being learned so that all fusion clients are learning the same function. Another approach to address data skew is to combine models using the concept of an ensemble. The ensemble can be coupled with policies to create an overall model that works better under a different set of conditions. The policies for ensembles can be generated so that they are picking a subset of models that are better trained for the environment under which the inference task is being done.

Chapter 6

Addressing Trust Issues in Federated Learning

While most scenarios for federated learning in enterprise settings are such that the fusion clients and fusion servers can trust each other, there are some scenarios under which the trust among the various parties would not be absolute. In these cases, the system for federated learning needs to take into account the restrictions that would arise due to the limited amount of trust among the parties. In this chapter, we look at some of these trust issues and approaches to address them.

When trust is limited, sites may be unwilling to share model parameters in the raw with other partners, and mechanisms will be needed where models can be built without necessarily sharing model parameters in the clear with parties that are not trusted. The need to not share model parameters in the clear could be a show-stopper in some contexts.

One way to understand these situations and to resolve them would be to use the concept of trust zones. A trust zone consists of systems that trust each other and are willing to share data and information with each other without restrictions. An example of a trust zone would be the servers within an enterprise firewall. Being protected by the firewall, the systems within the enterprise trust each other more than systems that are outside the firewall. In larger enterprises, in data centers hosting multiple applications, or in cloud hosted locations, there may be multiple trust zones protected by a series of multiple firewalls or other security devices.

Typically, in the rest of the discussion about federated learning in this book, we have been assuming that all systems belong to the same trust zone, i.e. the different fusion clients and the fusion server all belong to the same trust zone. In this chapter, we examine the situations which include more than one trust zone and discuss approaches to enable federated learning across trust-zones.

6.1 Scenarios with Multiple Trust Zones

Some scenarios where federated learning needs to work across multiple trust zones include the use of cloud based services, consortia, alliances, and military coalitions.

6.1.1 *Cloud-based Fusion Server*

Cloud computing [21] has become one of the dominant modes of providing services and computing infrastructure. From the discussions of Chapter 4 and Chapter 5, it would be clear that the fusion clients and fusion servers need to implement a complex pipeline of functions to enable the entire system of federated learning to work together. One way to simplify the complexity is to run the fusion server as a service available for use as a cloud-hosted service.

However, when the fusion server is run and operated as a cloud-hosted service, and the data and fusion client is present within different branches of a company, trust between the fusion server and the fusion clients may be limited. Consider a bank which may have its local data warehouses at several different locations, and would like to build a common AI model across the data in all of the warehouses. A cloud provider may offer a fusion server as a cloud service to the bank. The bank may be interested in leveraging the fusion server in the cloud, since it would save time in getting to a good AI model for use within a business process. However, the bank may be worried about the transmission of the data or even the model to the cloud service, and any security exposure or information leakage from the model.

The situation is shown in Figure 6.1. Using the fusion server in the cloud would provide an enterprise with many useful utilities, which can address the issues with data format differences and data skew discussed in Chapter 4 and Chapter 5. The cloud service would include the ability to define policies for data quality, policies for transforming data into a common format, identification of missing labels and various other functions. While the bank can replicate these services within own computing environment, it may find it more convenient, expedient and cost-effective to just leverage these functions as a packaged service provided by a cloud service provider.

One could argue that the fusion server in the cloud never gets to see the raw data of the bank, and it only has visibility to the model parameters. That level of protection may be acceptable to some banks. However, some banks may not be comfortable even sending models to the cloud service provider.

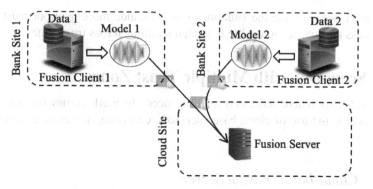

Figure 6.1: Cloud-based federation scenario.

There have been several papers that have shown how information can leak from a trained model. One could perform a model inversion attack on machine learning algorithms. Model inversion attacks allow someone to determine the input that was used to train a model. One approach to perform a model inversion attack is by examination of the confidence score of selected test samples passed through a trained model. This has been used to reconstruct faces provided for training of face recognition models [76].

Furthermore, neural networks can be trained to discriminate between samples belonging to a data set and those that do not using techniques such as Generative Adverserial Networks or GANs [77]. When used maliciously, GANs can be used to recreate parts of the training data set [78]. This means that the cloud service provider would be able to recreate some of the training data sets, which would be an undesirable information leak.

It should be noted that almost all of the attacks and information leaks published in the scientific literature can happen only under some rather carefully crafted conditions. The normal business processes and operating practices of the cloud service provider would typically offer a strong protection against deliberate data leakage. Thus, the concerns about information leakage may be purely academic as opposed to a real concern in practice. However, banks may be concerned about possible compromise in the security of the cloud service provider. Even though the probability may be small, the banks may still be skittish about sending the models to the cloud service provider in the clear and be concerned about the leakage of their data.

In this particular scenario, the different fusion clients belonging to the bank are in one trust zone, whereas the fusion server in the cloud is in a different trust zone. In order to handle this situation, we need a solution for federated learning where the models do not cross trust zones in the clear.

6.1.2 *Multi-tenancy Cloud Site*

One of the environments where we pointed out that federated learning would be very useful was in the context of sharing information about different tenants that may be hosted by the same cloud service provider, as described in Section 2.4.3 in Chapter 2. Specifically, if the cloud service provider is managing applications for many different tenants, it would be able to train AI models based on the errors it may see in the logs produced by the hosted applications of different tenants. These logs could help the cloud hosted provider get early warnings about potential failures of the hosted application, and it would be able to leverage and combine the information across several tenants in order to improve its ability to proactively take actions against possible errors. The situation is shown in Figure 6.2.

As shown in the Figure, the cloud service provider is careful to not move any raw data or logs out of the firewalls that protect the servers belonging to individual clients, client 1 or client 2. Only the models are moved out of the client environment, protected by the client firewall, and the fusion server, located in the management site of the cloud service provider that is used to combine the models. The service provider could assure each client that their application logs, or any information about the management of their servers, is kept only within their enclave of the cloud.

While the creation of such models which can help the operation of a hosted service and improve the operation of the cloud service provider are very promising, it also causes a potential cause for concern among the clients that are being serviced by the cloud service provider. The clients that the cloud service provider is hosting may not like the fact that models trained on their data are being mixed with models trained on data from their competition. Their concern may not be that of any data leakage to the cloud service provider. After all, they trusted the service provider to run their servers and applications for them. However, they may be concerned about the use of the data from other clients hosted in the same cloud

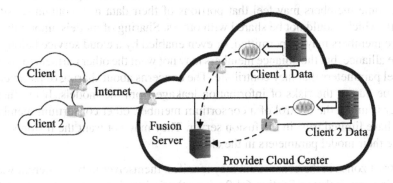

Figure 6.2: Multi-tenancy federation scenario.

environment. They may not want the other sites to have any information gleaned about their data that is being used to create the shared model.

The trust relationships among the different parties of the federation system is very different in these cases. The fusion clients trust the fusion server. However, the different fusion clients do not trust each other.

6.1.3 *Consortia and Alliances*

In many industries, consortia are formed to enable the sharing of knowledge and information among different organizations. As an example, it is not uncommon for health-care organizations of different countries [28, 29] to form an alliance to share information about health-care data with each other. Several joint organizations have been formed to share information about agriculture, social issues, and demographics among member organizations.

Alliances may also be formed among different academic, industrial and government organizations to conduct collaborative research. A few examples of these include the International Technology Alliance in Network Sciences [79] , the International Technology Alliance in Distributed Analytics and Information Sciences [80], robotics collaborative research alliance [81], etc. All of these alliances bring together researchers from different organizations, and share research activities with each other. Members of the research program collaborate by sharing data, knowledge and experience to solve a joint problem. While research organizations work together, they may not always have complete freedom of sharing data or models with each other. Industrial members would have additional restrictions placed upon participation due to the commercial interests of the company. Government members would have additional restrictions placed on their operation due to the regulatory requirements they need to satisfy. This creates a situation with limited trust among different parties.

Consortia are formed to share information among different organizations. However, some members may feel that portions of their data may contain sensitive details which should not be shared with others. Sharing of models among the alliance members may be permitted, or even enabled by a cloud service belonging to the alliance, but the alliance members may not want the other partners to see the model parameters that they contribute. The concerns about sharing of parameters may be due to the risks of information leakage from the models. If the fusion server is under the control of a consortium member, other consortium members may have limited trust in the fusion server. They may not want the fusion server to see their model parameters in the clear.

The trust zones in the consortium may manifest themselves in two different ways. The first way is that of Section 6.1.2 where the fusion clients may be comfortable sharing the model parameters in the raw with the fusion server hosted by the consortium, but not with fusion clients belonging to other members. The other way would be when a group of consortium members may trust each other, but do not completely trust a third consortium member or a group of other consortium members.

6.1.4 *Military Coalitions*

Modern military operations are frequently conducted within coalitions [82], where more than one country joins forces to conduct missions in a collaborative manner. Such coalition operations are a common occurrence in peace-keeping, where multiple countries work together to maintain peace in a region subject to conflict. They are also used for humanitarian operations in the event of natural disasters.

Coalition operations have their own set of technical challenges in multiple areas that need to be explored [83, 84], including challenges in federated learning [69]. As coalitions work together, almost every member nation collects data during the conduct of its operations. Such data may consist of video footage from surveillance, audio recordings and seismic readings from sensors, or tabular data that is maintained manually by the personnel involved in the joint effort. Much of this data can be shared among coalition members to improve their operations. As an example, the video footage of insurgents collected by the member nations can be shared to improve the AI models used to detect insurgency and to alert the peace-keepers.

Coalition members may not be able to share raw data, but may be able to share model parameters with each other. One specific case which that may arise is shown in Figure 6.3, where one of the coalition members is training the AI model but using the information available from all of the coalition partners. Each coalition partner would be willing to share the models, but not if the model

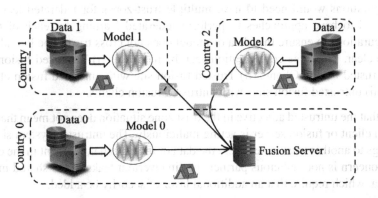

Figure 6.3: Coalition federation scenario.

parameters are sent in the clear. The trust zone for federation server includes the federation agent that belongs to the same nation, but each of the other agents are in their own trust zones.

A situation analogous to coalition operations can also arise in other emergency situations where different government agencies are cooperating together in any joint operation, e.g. reacting to a natural disaster, or planning for a special event. Different types of regulations may prevent complete sharing of raw data or un-protected model parameters among the agencies, but they would be able to train models and share the models with each other, especially if the model parameters can be encrypted during the fusion process.

6.2   Trust Zone Configurations

From the discussion of the different scenarios in Section 6.1, we can identify four possible configurations of trust relationships that occur in the different settings of federated learning. These four possible configurations are shown in Table 6.1 and describe the relationship from the perspective of a single fusion client to that of fusion server and other fusion clients.

Table 6.1: Possible trust configurations.

Number	Fusion Server	Other Fusion Client
1	Trusted	Trusted
2	Trusted	Untrusted
3	Untrusted	Trusted
4	Untrusted	Untrusted

The case where everyone is trusted requires only one trust zone. The three other configurations would need to have multiple trust-zones for federated learning. We need to have approaches to perform federated learning for each of these configurations to ensure that data or model does not cross trust-zone boundaries in the clear. The three configurations can be named as (i) Untrusted fusion site with trusted Fusion Clients (ii) Trusted fusion site with untrusted fusion clients and (iii) untrusted fusion sites with untrusted fusion clients.

Note that the untrusted adjective in the trust zone situation does not mean that any fusion client or fusion server is acting maliciously. The untrusted system simply belongs to another trust zone where the data or model can not be sent in the clear. The concern is not malicious partners but inadvertent leaks of the shared model or data, which require that an additional layer of security be added.

6.2.1 *Trusted Fusion Server with Untrusted Fusion Clients*

In this configuration, shown in Figure 6.4, each fusion client trusts the fusion server and is willing to share the model and other statistics with the server. However, fusion clients do not trust each other.

For this configuration, there are many different trust zones, each trust zone including a fusion client and the fusion server. The fusion server is tasked with the challenge that peer clients not be able to obtain information about the models provided by any other individual fusion client. Note that, in this particular situation, the fusion server belongs to all the trust zones, and has the responsibility of ensuring that data or model does not cross between the trust zones.

This configuration would arise in the scenarios described in Section 6.1.2. Note that this may also occur in some consortium scenarios (Section 6.1.3).

A consortium may be formed to provide specific services to each individual member. While individual members are willing to trust the consortium (established as an independent entity), they may not necessarily have the same level of trust with other members of the consortium. An example of such a consortium would be the Federal National Mortgage Association (FNMA) also known as Fannie Mae. It provides common services to banks and financial firms that are in the business of offering mortgage loans in the United States. Fannie Mae is supported by the U.S. government and operates as an independent company. If Fannie Mae were to host a fusion server to provide AI model building services to support a group of small banks to improve their mortgage processes, each bank would have a level of trust in providing its models to Fannie Mae. However, the banks may want assurance from Fannie Mae that their models would not be revealed to competing banks.

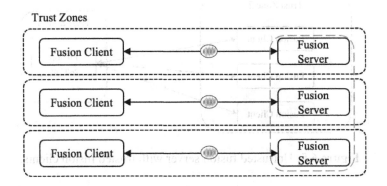

Figure 6.4: Trusted fusion server with mutually untrusted fusion clients.

6.2.2 *Untrusted Fusion Server with Trusted Fusion Clients*

In this configuration involving multiple trust zones, the fusion clients trust each other, but do not have a complete trust in the fusion server. This trust configuration is typical for the cloud-hosted scenario described in Section 6.1.1. A similar situation can also arise in coalition operations if several sub-units from one country are leveraging the infrastructure of another country's military to improve their AI models. In a military coalition involving one country which has a significant infrastructure for operations in some geography assisting a country which may not be technology advanced, this situation may arise. An example would be a coalition led by the United States to help the fictional country of Gao in the hypothetical coalition scenario of Binni [85] or the United States helping the fictional country of Holistan in the Holistan scenario [86]. In these scenarios, the United States would be operating bases to help out in peace-keeping operations in areas of the host country (Gao or Holistan). US forces would have data at many locations, i.e. it would have fusion clients in its different bases and may be leveraging the fusion server hosted by the host country to create a composite model.

The untrusted fusion server with trusted fusion clients configuration is shown in Figure 6.5. The different fusion clients belong to the same trust zone which is marked as trust zone 2. The fusion server, however, belongs to a different trust zone, which is marked as trust zone 1. Between the two trust zones, models and data summaries are being exchanged. In order to perform tasks such as model fusion and other policy negotiations, the fusion server needs to operate in a manner without seeing the model parameters in the clear.

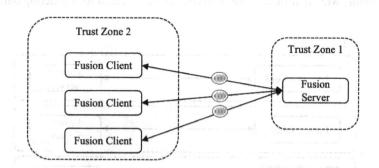

Figure 6.5: Untrusted fusion server with trusted fusion clients.

6.2.3 *Untrusted Fusion Server with Untrusted Fusion Clients*

This configuration would arise in the environments where the fusion clients do not trust the fusion server, nor do they trust the peer fusion clients. This configuration in shown in Figure 6.6. Each fusion client and the fusion server are in different

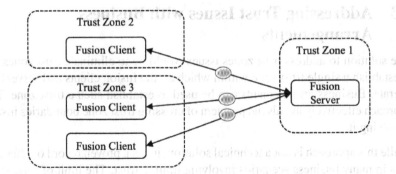

Figure 6.6: Untrusted fusion server with mutually untrusted fusion clients.

trust zones. Note that some fusion clients may trust each other, but we can then cluster them together as being in the same trust zone. In the figure, the trust zone of the server is marked as number 1 and the client trust zones are numbered from 2 onward. Fusion Clients want to protect their models from other clients, as well as the fusion server.

This configuration may arise in consortium, alliances and military coalition scenarios. In a military coalition among several nations that all have access to advanced technology, one of the allies may offer to run the fusion server. The other allies in the coalition would be running their fusion clients, but they would have limited trust with either other fusion clients, or with other fusion servers.

In a consortium formed among different companies to share health-care or demographic information, a similar level of limited trust scenario may arise. Each member would be operating their fusion client, and would have limited trust in the fusion server. Similarly, they would not be willing to trust other fusion clients that are being operated by other companies.

The different scenarios described in Section 6.1 can be mapped to different configurations with multiple trust-zones, as shown in Table 6.2.

Table 6.2: Mapping of scenarios to trust configurations.

Scenario	Trust Zone Configuration
Cloud-based Fusion Server	Untrusted Fusion Server with Trusted Fusion Clients
Multi-tenancy Cloud Site	Trusted Fusion Server with Untrusted Fusion Clients
Consortia and Alliances	Untrusted Fusion Server with Untrusted Fusion Clients
Military Coalitions	Untrusted Fusion Server with Untrusted Fusion Clients

In the next few sections, we would discuss approaches to perform federated learning under these different configurations of trust zones.

6.3 Addressing Trust Issues with Business Arrangements

One solution to address trust zones issues would be to eliminate trust zones and to establish a single trust zone within which all the fusion clients and servers can operate. Business arrangements can be used to establish such a trust zone. This approach effectively avoids the problem of crossing trust zone boundaries instead of solving it.

While this approach is not a technical solution, it has a proven record of effectiveness in many business scenarios involving limited trust. The limit of trust arises due to the fact that different entities in the federated learning process, namely the fusion clients and the fusion servers, belong to different organizations. The primary concern of the organization with sharing data or model with other organizations is the possibility of information leakage. If any sensitive data leaks out, there may be financial implications, which may arise due to violations of regulatory requirements, the costs incurred for mitigating actions where a data leak happens, or due to the loss of business due to bad publicity that arises due to data leakage.

In each of the three configurations of trust zones identified in Section 6.2, the primary interaction happens between the fusion client and the fusion server. While it is feasible to perform any task performed in the client-server manner in an approach which is peer-to-peer without involving any clients, the complexity of performing that is much more difficult than that of the client-server architecture. Therefore, we will assume that the interaction and business arrangements happen in a purely client-server manner.

The business arrangement in these cases would typically consist of a contract between the client and the server in which each organization would specify the manner in which their data will be handled, and any financial penalties that may be levied by either party in case the agreed upon manner is not followed properly. The two organizations would agree to share sufficient information to check on each other and ensure that the proper adherence to the agreement is being made. Once those arrangement are made, the models and data may be exchanged in the clear across the trust zones.

This type of arrangement is established practice among many businesses, and while not a technical solution, often provides the most expedient approach for working across different trust zones. A common term used for such arrangements is a service level agreement, which can be defined for various aspects of the security of data and any other digital content (e.g. models) that are to be shared among partners. Service level agreements have been defined for various types of arrangements, e.g. security [87], cloud hosted services [88, 89] and computer communications networks [90]. The same type of agreements can be provided and arranged for in federated learning.

6.4 Addressing Trust Issues with Infrastructure Technology

While business arrangements and service level agreements provide an incentive for people working within an organization to mitigate any of the issues that arise during the task of building data or sharing models, they would need to be complemented by creating an infrastructure for training models and sharing information which would minimize the likelihood that the data or models are being compromised. There are many infrastructure technologies, and they are all targeted at ensuring that different issues that are outlined within the contractual agreements among the businesses are maintained.

The traditional security mechanisms for data and model management need to be followed to ensure compliance with the objectives that are specified within the business arrangements. These may include provisions such as the data statistics and model parameters that are exchanged by the system and not stored in the file system, that they be maintained only within the memory or volatile storage of the system and that all copies of the statistics and model be destroyed once the task of federated learning is over. They may also stipulate that access to any computers that are involved in the task of federated learning be restricted to authorized personnel.

Some enterprises may opt to participate in federated learning only by carving out special zones within their IT infrastructure which are designed to be accessed by people outside the enterprise. It is not uncommon for enterprises to classify their IT infrastructure into different zones which may carry designations like the red zone, yellow zone, green zone and blue zone. The red zone would consist of computers that are allowed to be accessible publicly, e.g. public facing web-servers. The yellow zone would consist of machines to which people not employed by the enterprise premises are allowed access, e.g. if there are arrangements for a partnership with another enterprise and some servers are made accessible only to selected employees of either the local enterprise or the partner enterprise. The green zone would consist of computers that are used by employees on-premises which are allowed access either by establishing a virtual private network, or by physical presence in the buildings owned by the enterprise. The blue zones may consist of computers that can only be accessed when present phyiscally in the building of the device, e.g. systems that include sensitive financial records of the company. All of these zones would be protected by a sequence of firewalls and other security devices, which allow limited communication back and forth among different computers and systems in the other security zones. There may

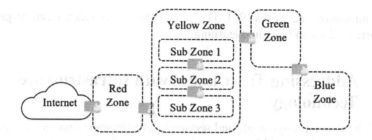

Figure 6.7: Security zones within an enterprise.

be other sub-zones within each of these security zones, e.g. if the enterprise has a partnership with multiple partners, it may have a sub-zone within the yellow-zone for each partnership arrangement, and may follow a different set of security protocols for each of the sub-zones. The structure of zones within a hypothetical enterprise is shown in Figure 6.7.

Different organizations and businesses may refer to the multiple security zones with different color codes, numbering schemes or names. However, the computer and communications infrastructure is frequently divided into these zones, with each zone provided mechanisms for security monitoring tools and appropriate administrative procedures to ensure that nothing untoward is happening within the enterprise.

When organizations that do not completely trust each other are involved in sharing models or summary statistics, they may choose to use infrastructure isolation technology to ensure that they are not subject to sensitive information leakage. The enterprise providing the data fusion services would place its servers into its yellow zone, uses its firewalls to ensure that the zone is only accessed by the identified servers from data generation sites that are participating in the partnership arrangement to share the data, and maintain the log of access to any servers accessing the data fusion server. The data generation sites themselves may also be putting any data they want to share, or any models built upon that data, into their own yellow (or equivalent) zone or sub-zone to ensure that they are protected from the fusion server. For each of these zones, they would also be deploying appropriate security monitoring devices and other measures to minimize the probability that any data or model component in the federated learning process is leaked.

A hypothetical situation where two banks are working with an agency (e.g. Fannie Mae) to create a shared model is shown in Figure 6.8. The banks do not trust each other since they are competing with each other. However, the banks trust the agency to share their models with the agency. However, since the trust is not absolute, the banks would put the data used for model building in their yellow

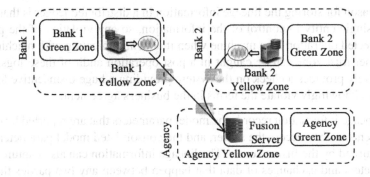

Figure 6.8: Example for infrastructure security approach.

zone, and share that with the agency using all the appropriate security mechanisms they have for the yellow zone. The agency similarly would have its own yellow zone and it would only put the data fusion server and the minimum necessary computational resources and data required to train a common model. Only the participating bank sites are allowed to access the fusion server provided by the agency. Both banks and agency would also be monitoring the yellow zone to make sure that their sites are used properly by the other participants and that no information leakage is happening.

6.5 Auditing and Logging

If data and models are going across different trust zones, technologies for auditing the data that is being sent, and proving that the data and models were handled properly is a requirement that can be often specified in business agreements. In order to share data and models, the different sites involved in federated learning must implement proper auditing and logging of requests and data flows within their sites. All requests that are made across the sites must be logged, and when audited by an independent party, should show that the proper procedures for handling the data and models coming from other partners and sites were followed.

One approach to track the evolution of models and data used to train them in federated learning contexts can be provided using blockchain [91]. Blockchain or distributed ledger technology [92] is technology that can maintain records about multiple transactions in a distributed ledger, which is not owned by any single organization, but is maintained using distributed consensus among a large group of peer to peer participants. These distributed ledgers can be kept to keep track of how data is used for training, and what model parameters are being exchanged across different trust-zones. By maintaining information about the federated learning model building process, one can maintain a lineage of the information used to create a shared model [91].

The reason for storing the lineage information in a distributed ledger is that there is no single entity in control of the information, so any violation of the proper or expected operation by a machine when it is communicating with machines in the other trust zone can be caught in a post-operation audit of these logs. This effectively provides a check on the system providing a huge disincentive for any party to knowingly violate the terms in the business agreement.

The lineage information records the model parameters that are reported by different clients as they access the server, and the consolidated model parameters that are returned by the fusion server. The lineage information can also capture other parameters and exchanges of data that happen between any two parties that are in different trust zones, including the recording of any parameters and policies exchanged during the pre-processing stages of federated learning, i.e. the various procedures described in Chapter 4.

Architecturally, the maintenance and storage of the logging operation is shown in a simplified manner in Figure 6.9. The example shown is the same one as in Figure 6.8 with two banks and an agency. The interactions in federated learning happen via interactions that happen on the sold black lines that invoke commands on servers in the different environments, requesting the appropriate functions to be performed. These interactions would typically go through the firewalls and other security devices that are not shown in the figure, but will be invariably present in any infrastructure. Additionally, each enterprise involved in the federated learning task operates a distributed ledger among themselves. One piece of the distributed ledger runs in each of the enterprises, and they all interact together to maintain a distributed ledger. The enterprises can record all information being sent or received in the joint operation with each other, and store it in the ledger for subsequent auditing. The distributed ledger protocols ensure that the records of these transactions are maintained in a manner that can not be forged or repudiated easily.

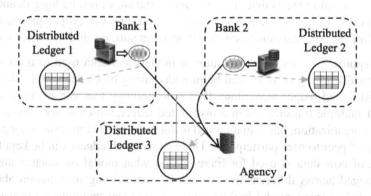

Figure 6.9: Distributed ledger-based logging.

Audits and logging technologies provide a mechanism for partners to validate and certify that they are adhering to the business agreements placed upon them in order to share data for distributed learning. It also provides a disincentive for anyone trying to manipulate the data or models offered in the process of federated learning. The one issue with maintaining lineage of all interactions happening in this manner is the overhead associated with recording all the information in a distributed ledger. The cost can be reduced by not logging every interaction that crosses trust-zones, but only a subset of the interactions. Alternatively, depending on the level of trust between the different parties, the logging and monitoring of lineage may be done by a third-party, who is trusted by all parties, and can keep track of the transactions happening between them.

The advantage of getting additional data for training needs to be balanced against the overhead and expense of maintaining the distributed lineage information, and a cost-benefit assessment needs to be done to determine which of the three courses of action is best from a business perspective, to perform federated learning without any logging, to not perform federated learning, or to perform federated learning with logging of lineage.

6.6   Encryption-based Approaches

The goal underlying business agreements, infrastructure security and auditing capability is to provide enough assurance to partners involved in the task of federated learning that there is minimal risk in exchanging models across the various trust zones. In many business contexts, they may be adequate to convince the partners to share their models and engage in a shared model training. Most businesses recognize that it is not possible to get absolute trust, the risk of information leakage must be weighed against the gains of building knowledge from additional data. As long as the risk of losing model information is lower than the business value gained by creating a shared model, many businesses would accept the information loss risk and exchange model parameters with each other in order to create a shared model. However, there may be some businesses which may not be willing to share models in the raw outside their trust zone, even with all of the procedures in place.

For such businesses, data and model sharing approaches based on encryption technologies might provide an acceptable solution. When using encryption, model parameters and even summary statistics about the data are not transferred in the clear across trust zones. Instead, only the encrypted forms are transferred across the trust zones. Encryption is the process which converts any piece of data (plain-text) into an alternative representation (cipher-text) so that only people who have access to some secret (keys) are able to extract the original plain-text from the cipher-text. In general, encryption destroys the structure of the information contained in the plain-text, and the cipher-text looks like randomized data. Sensitive information exchanged over the Internet is usually encrypted.

Modern encryption techniques on the Internet usually follow a combination of two techniques (i) symmetric key-based encryption and (ii) public key-based encryption. In symmetric key-based encryption, both parties involved in a communication know the secret key. The sender uses the secret key to transform the origin text to cipher-text and the receiver uses the same secret key to do the inverse transformation of cipher-text to plain-text. In public key-based scheme, each party has a pair of keys, one key is kept secret (the private key) and the other key (public key) is made available to anyone wanting to communicate with the party. The key-pairs are generated so that text encrypted with private key can be transformed to plain-text with the public key. The sender would encrypt the plain-text with the public key of the receiver, and the receiver does the inverse transformation using the secret private key. Public keys are made available usually in standard digital representations called certificates which contain the public key associated with a party. The certificates are signed with the private keys of some trusted sites whose public keys are well-known to everyone. In Internet-based communication, a client would use a certificate to obtain the public key of the server, use public key cryptography to negotiate a secret key with the server, and use the secret key to encrypt and decrypt messages for some time. The secret key can be refreshed at periodic intervals. Since public key-based encryption and decryption is more computationally expensive than symmetric key encryption, this mechanism provides a good secure efficient scheme for secure communication. Surveys of encryption approaches can be found in references [93, 94].

When using traditional encryption algorithms, the cipher-text becomes close to randomized representation of the plain-text, and can not be processed. However, new forms of encryption algorithms have been proposed which allow operations like addition and multiplication to be performed on encrypted cipher-text. This class of encryption algorithms exhibit homomorphism, which is defined as the ability to preserve relationships present in the original form of a data in the transformed (in this case encrypted) form of data. This class of encryption algorithms, called homomorphic encryption algorithms, can allow the sharing of models and data with other parties that can perform operations on them without revealing the plain text. Two categories of homomoprhic encryption schemes exist, known as fully homomoprhic encryption and partial homomorphic encryption, as described in the following subsections.

6.6.1 *Fully Homomorphic Encryption*

A fully homomorphic encryption scheme is an encption scheme which satisfies the following two properties:

- Additive Homomorphy: The encrypted version of sum of two plain-texts is the sum of the encrypted versions of two plain-text values

- Multiplicative Homomorphy: The encrypted version of the product of two plain-texts is the product of the encrypted versions of two plain-text values

Expressed in an alternative version, supposed x and y are two numbers and their encrypted versions are $E(x)$ and $E(y)$. Then the two properties are:

- Additive Homomorphy: $E(x + y) = E(x) + E(y)$

- Multiplicative Homomorphy: $E(x.y) = E(x).E(y)$

An encryption algorithm that satisfies both of these homomoprhic properties is known as a fully homomorphic encryption algorithm. The first fully homomorphic encryption algorithm was proposed in 2009 [95], with subsequent research leading to a couple of other encryption algorithms satisfying the property [96, 97].

The development of such algorithms means that any operation that can be performed on the data in the clear can also be done on data that is encrypted. Any algorithmic operation can be performed as a combination of various addition and multiplication operations. As a result, fully homomorphic algorithms enable computation on data that is encrypted. This allows an enterprise to send data to a cloud server to perform complex processing on it without revealing the data to the cloud server. This capability allows partners involved in federated learning to exchange model parameters and data statistics in an encrypted format.

The primary challenge associated with fully homomoprhic encryption is the performance hit it produces within the system. The standard implementations of an operation running using these algorithms ran 14 orders of magnitude slower (i.e. 10^{14} or 100 trillion times slower) than the plain-text operation. Since then, several researchers have been proposing enhancements to the algorithmic implementations using hardware accelerators and parallel processing [98, 99], which can reduce the encrypted operation to work within 2 orders of magnitude (2 times slower). However, that is still a significant overhead.

In future, schemes exploiting fully homomoprhic encryption would be viable as more enhancements in their implementation are achieved.

6.6.2 Partial Homomorphic Model Learning

In a partial homomorphic algorithm, only one of the two homomoprhic properties is preserved, i.e. the encryption algorithm either obeys the additive homomoprhic property or it obeys only the multiplicative homomoprhic property. Several algorithms proposed for encryption show one of these properties. The RSA algorithm [100], which was one of the first public key encryption algorithms, exhibits multiplicative homomorphism. The El Gamal algorithm [101] is another popular public key encryption algorithm that exhibits multiplicative homomorphism. The Paillier algorithm [102] shows additive homomorphic property. The advantage of the partial homomorphic algorithms is that their performance overhead in encryption and decryption is perfectly acceptable in the normal conduct of business and these algorithms are used within current applications and implementations.

Since only one of the operations, either addition or multiplication can be done on encrypted content while preserving the relationship, partially homomorphic encryption can be used to enable a form of federated learning where the operations performed by the federation server are done on encrypted content. This approach has been proposed to address configurations where the fusion server is untrusted [103]. The tasks that are to be performed by the fusion server are restructured so that they only need to perform one type of application, either multiplicative or additive. If the encryption algorithm shows additive homomorphy, then the operations are structured to be addition only. If the encryption algorithm shows multiplicative homomorphy, then the operations are structured to be multiplication only.

For example, if an algorithm using the federated averaging of parameters of a neural network as described in Section 3.4, the federation server performs the task of summing up the model parameters and then dividing them by the number of participants to get the average. This operation includes both addition and multiplication (by $1/N$ where N is number of participants). The operation can be structured so that only the addition is performed at the server, and given the number of sites participating in the federated learning process, the multiplication operation can be done by each of the sites themselves. Assuming an additive multiplication encryption algorithm is being used, the fusion clients can send the model to the fusion server in an encrypted format, and the server can perform the operation of addition on the encrypted content. If a multiplicative homomorphic algorithm is used, the exponents of parameters would need to be sent, since the results of multiplying the encrypted components, and taking the logarithm at the clients would result in the addition operation.

In order for the scheme to work, the keys for encryption have to be negotiated so that all the fusion clients use the same encryption key. This key can be provided by a server in the trust zone of the clients, or by means of electing one of the client sites as the key generator. The server in the other trust zone can help in this election, or the first fusion client can automatically become the key server for all other clients that join subsequently.

When the trust zones are different, partial homomoprhy requires that the operation to be performed be reformulated in a manner so that it is done only using all additions or all multiplication operations when the computation needs to be sent to an entity outside the trust zone. In some cases, this can add a significant amount of complexity to the task, and sometimes may not even be feasible. For those trust zones where this reformulation is possible, partial homomorphic encryption can provide a very viable approach.

6.7 Differential Privacy-based Approaches

In trust zone configurations where the clients do not mutually trust each other, they may not want to send the model parameters to the server to be shared with other clients. When there is only one other client, such sharing can reveal the parameters to the other client. Similarly, if the other clients are colluding together, they can determine the model parameters of the client. If it is important for a fusion client to not expose its model parameters to the other clients, it can use introduce some noise to the model parameters, but add the noise in a manner which does not impact the final outcome of the neural network being built.

The general area which discusses how the noise ought to be explored is the field of differential privacy [104], which looks at various statistical properties of the system, and how the noise can be introduced in an intelligent manner. Instead of going into the complexities of all aspects of differential privacy, we would simply illustrate the problem with the case where statistical property of the system, e.g. the mean and variance of the range of a feature need to be determined across the data held at all of the sites. This would be required to obtain a consistent scaling of data at all sites to a mean of zero and variance of 1, as described in the approaches described in Section 4.1.2 for ensuring consistent scaling across all of the sites. However, any given site does not want others to know of its individual average value or the variance, but they all want to learn about the mean and variance values across all of the sites together.

If the sites trusted each other, each fusion client site could report its mean and variance along with the number of data points it has. The fusion server would compute the weighted average of the supplied mean and variances to all of the parties. However, if a client does not trust the server, and is concerned about the leakage of information to other clients or the servers, it may not want to report its data in the clear. Noise addition allows the calculation of such values across all the sites without the information leakage.

In order to obtain the weighted average in a differentially private manner, the sites would participate in two rounds of a differentially private averaging process. In the first part, they would each find out the average number of points across all the sites. This allows them to get the sum total of the data points at all sites, by multiplying this average with the number of participating sites. In the next round, the sites would compute the average of the weighted means of their sites, i.e. they multiply the average of the values they have with their fraction of the overall count of data points, and calculate the differential private mean of the same.

When the trust configuration is that of untrusted fusion server and untrusted fusion clients, one way to calculate the overall differential private mean is for each site to introduce some error within their mean that they report to the fusion site. However, there is coordination in how the noise is introduced, e.g. each site has agreed that they would introduce a noise that is taken from a Gaussian distribution with a mean of zero and a common variance which is agreed upon by all of the fusion client sites. If each site adds a random noise to their mean value based on the distribution, then the averaging of the noise across all of the sites is likely to cancel each other out, assuming that the number of sites is reasonable. With N fusion sites, the mean noise added cumulatively would be expected to cancel out with a variance that is $1/N$ times the variance that everyone chose to use. By iterating over the process a few times, the cumulative mean can be calculated, which is highly likely to be accurate even though none of the sites ever sent out their true mean to the fusion site. Similarly, the other fusion clients can not really guess the correct mean value at any of the other site.

Beyond this simple example, statistical techniques in differential privacy provide mechanisms to add noise so that other types of properties about the distribution of any parameter (e.g. maximum value, minimum value, percentiles, etc.) can be calculated without sharing the correct information from any of the sites. This allows for a way to protect the model parameters, and the summary statistics information while participating in the fusion process. Noise addition approaches would have much less overhead than encryption based approaches, and may be acceptable for specific trust relationships.

6.8 Summary

In general, trust issues among businesses can be solved using business arrangements and appropriate mechanisms for infrastructure security, and agreement among the parties for auditing and logging of the data that is shared or exchanged. When those mechanisms are not acceptable, data hiding mechanisms, including use of homomorphic encryption and differential privacy techniques, can be used.

Chapter 7

Addressing Synchronization Issues in Federated Learning

One of the assumptions made in the Naive federated learning algorithms described in Chapter 3 is that different fusion clients are running concurrently in a synchronized manner with each other. This concurrent operation simplifies the operation of a fusion server. However, from logistics and operational perspectives, concurrent training introduces a significant amount of complexity and reduces the viability of a federated learning solution. A practical federated learning solution must allow fusion clients to be active at different times at a schedule which is determined by the fusion client, as opposed to operating under the control of a remote fusion server.

In the first section of this chapter, we discuss the issues associated with synchronization. The other sections of the chapter deal with some approaches that can be used to address those issues and deal with situations where concurrent operation is not feasible.

7.1   Overview of Synchronization Issues

To illustrate the issues with synchronization, let us assume that there are some N fusion clients and one fusion server. In order for the naive algorithms of Chapter 3 to work, the assumption is that all the fusion clients and the fusion server are active concurrently. A time-diagram showing when different fusion clients and servers need to be active will be similar to that shown in Figure 7.1. All the fusion sites and fusion servers need to be active and running at the same time.

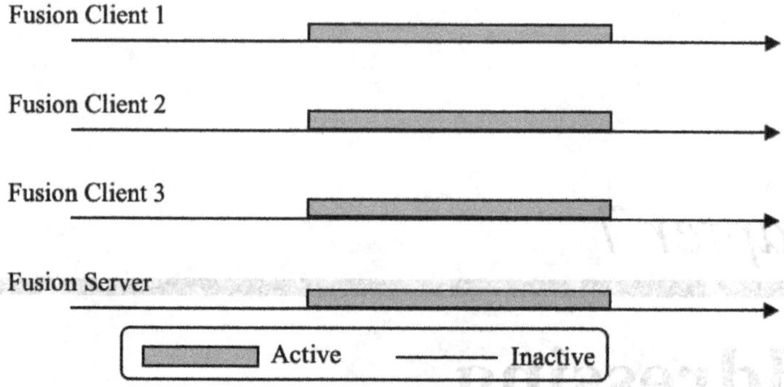

Figure 7.1: Activation diagram for naive federated learning.

The fusion clients at all of the sites need to be physically started at approximately the same time that the fusion server is started. A little bit of staggering in the starting time may be permissible as different clients join the servers, but the server needs to wait till all the clients are connected. Once the clients are all connected to the servers, they can go through the procedures required to address data mismatch and data skew issues and train the models together.

While this synchronization may appear to be a trivial requirement, it can cause a significant challenge for implementation in a distributed environment, specially when the different sites are under different administrative controls. Considering the physical layout of the federated learning environment, different fusion client sites may be in different countries, or in different cities of a large country. If they are located within an industrial or government facility, they are likely secured by security devices which ensure that the data does not travel outside their facilities, prevent inbound connection requests and block the majority of traffic, requiring explicit permissions for any allowed network communication.

The operational complexity arises due to the need to start multiple fusion clients in different locations at the same time. One option would be to have an administrator at each site start the program to run at a specific time. The administrator can run the program at the prearranged start time, or can instruct the server to automatically start the program at the specific time. Regardless of how this coordination is performed, it is a manual step which requires a significant bit of overhead, and it is possible that mistakes made by the administrators can mess up the process by not starting some of the services required for this operation. Even with automated scripts, some clients may fail to start holding up the model building process. Furthermore, if the time needs to changed, e.g. if the scheduled time for model learning coincided with a scheduled down-time of the fusion server, the adjustment of the scripts for modified time needs to be coordinated all over again.

A better option would be for each fusion client to provide a mechanism through which a remote site, e.g. the fusion server site, could start all the fusion clients at the appropriate time. If the fusion client sites permitted such types of inbound invocation, synchronized activation would be trivial for a program to initiate. Instead of having a distributed team of humans, a single human can start the federated learning process from the fusion server. The operational challenge here comes from the security needs of the enterprise. In order to have better security, most enterprise sites restrict remote users from accessing a server on their location. Machines can connect out to the servers, e.g. fusion clients can initiate a remote connection to the fusion server, but fusion clients are not allowed to accept connection requests from outside the firewall. If the fusion server were operating as a cloud-hosted service, the need to adjust permissions on fusion clients would be a significant impediment in the use of the cloud service.

Such a coordination challenge will arise if a consortium were trying to train a common model using federated learning. Let us assume that all members of the coalition have all collected data and maintain it in a manner so that a common model can be built. They all have the resources and servers on site to train the models and exchange them with the fusion server, which is hosted in a cloud service owned by the consortium. Each company has followed the security practices of their locations, as described in Chapter 6 and put the data and machines running the fusion client into their appropriate security zone. In order for a data scientist running the fusion server hosted in the cloud to invoke the fusion clients to start the training process synchronously, each company has to allow the inbound call from the cloud service to their fusion client. This type of external service invocation on a server is usually difficult (although not impossible) to obtain in many companies, given their legitimate concerns about the security exposures of systems that are accessible over the Internet. Most enterprise security teams would prefer to only allow outbound calls from the servers within their control.

The problem is illustrated from the perspective of a single enterprise running its fusion client in Figure 7.2. The enterprise has placed its data and fusion client so that it can access the fusion server running in an external cloud service. The enterprise does not want its data to be revealed to the fusion server, or to other parties, which is why it is engaging in federated learning instead of simply transferring the data over to the site for fusion. It would deploy security devices to protect its data. It is very common in the enterprise to enforce security by only allowing outbound connections from its locations, and to restrict any inbound connections on servers it hosts to enterprise premises. Therefore, the enterprise would be allowing outbound connection requests from its premises to the cloud service, but rejecting any inbound connection attempts. This makes the task of synchronization from the fusion server difficult, since the fusion server can not invoke the fusion client in the enterprise to initiate its operations.

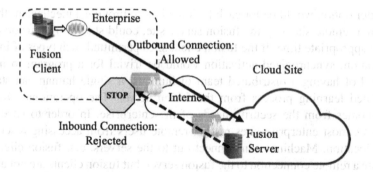

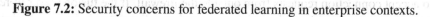

Figure 7.2: Security concerns for federated learning in enterprise contexts.

A similar challenge arises if a service provider company, like Fannie Mae, were to offer a service to build an improved mortgage risk prediction model for banks. Being a distinct company, even though trusted by banks, Fannie Mae would have a hard time ensuring that all bank fusion clients are ready and operational at the same time so the appropriate models can be built. It is possible to have that time of scheduled operation, but it would be significantly easier if banks could simply upload their trained models to the Fannie Mae Fusion Server at their own convenience.

There are techniques which can allow synchronization even when inbound connection is not allowed. One can have the fusion client maintain a periodic request to the fusion server to check whether it is time to start the active training process. Although somewhat inefficient, it does allow for an approach to synchronize the operations, once the fusion server knows that all fusion clients are available for a synchronized training session to being. Another option is to use a protocol like websocket [105] which would enable messages to be sent back and forth between the client and the server and allow the synchronization to happen. This protocol traverses the firewall over the ports used for web traffic, and effectively implements the synchronization mechanism of period checking within the protocol implementation itself. A third approach which does not use technology, but depends on business arrangements, is to implement the proper security, auditing and logging mechanisms so that the fusion server is allowed to make inbound calls.

Even if the proper arrangements to allow synchronized operations of different clients and servers are implemented, a problem that still remains is the change in the membership of the consortium over time. The situation is shown in Figure 7.3. The consortium begins originally with 2 members, and both members can train a shared model using federated learning algorithms. The models provided by the two companies are M_1 and M_2, which are combined by the fusion server into a model C_0. A few months after the consortium is formed, a new company wants to

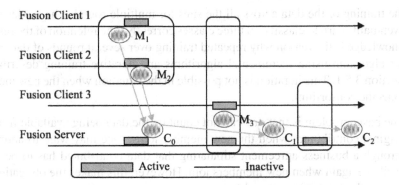

Figure 7.3: Group changes causing synchronization issues.

join the consortium. The third company would contribute a model M_3 trained on its own data. However, when the model is available from the third company, the other two companies may have decommissioned the infrastructure they set up for sharing of the data with the consortium. The consortium leader has to augment model C_0 with the new model M_3 provided by the new company, but the data from the original two partners is no longer available. An approach which relied on having everyone do the training together at the same time would not work. Instead, an approach which can combine the models C_0 and M_3 to produce a new model C_1 is needed.

It may appear that taking model C_0 and asking the member to train it further on their data might be a way to address the problem. Unfortunately that does not work in most of the cases. Many popular model training algorithms, particularly the common ones used for deep neural networks, suffer from a problem called catastrophic forgetting [106]. When an already trained neural network is trained with new data, they learn the new patterns but forget the patterns that they had learned earlier.

Catastrophic forgetting is one of the key issues which renders federated learning difficult when data is skewed or partitioned [71] as described in Chapter 5. Suppose there are three sites, each with data that belongs to only one class, and each of the sites has data belonging to a different class. When the model is trained on data from site 1, it would only predict each input as belonging to class 1. When this model is then incrementally trained on data belonging to site 2, one would expect it to be able to distinguish between class 1 and class 2. However, the model forgets the characteristics of class 1 and, after training over the data from site 2, would overwhelmingly predict the output as belonging to class 2. Similarly, the model would overwhelmingly predict any input as belonging to model 3 after training at site 3, largely forgetting any contributions made by either of the two previous sites. The forgetting is not complete, and if one keeps on iterating on the training of the data across all the sites for multiple rounds, the model would eventually learn to classify all three classes correctly. The retention of the residual knowledge is the reason why repeated training over several rounds of data works for algorithm using incremental algorithms and iterative training described in Section 3.5.1. This iteration is not possible in the situation when the new member joins the consortium.

The basic problem is that one can not count on the data being available from the original contributors when the new member joins. One may want to address it through a business agreement stipulating that data contributed has to be made available again when new members join. However, this makes the obligations of joining the consortium more burdensome.

Another problem can arise if one of the members of the consortium decides to leave, and insists that the patterns from its data be removed from the consortium model. The consortium now has the challenge of creating a model which did not include any data from the departing member to change its current model C_1 into a model C_2 which does not include any part of the model M_1, but is only the result of combining model M_2 and model M_3. This problem can not be addressed by the algorithms we have discussed in Chapter 3.

If the consortium is built one partner at a time, a situation can arise where each partner is providing their models one at a time. This would require the ability to compose models one at a time without any synchronization among them.

7.2 Asynchronous Data Mismatch Issues

The approaches to handle data mismatch described in Chapter 4 assumed that all the fusion sites were active. It allowed the sites to exchange information with each other about the format of data that they had, and the best format to use for the model training could be determined. Similarly, in order to scale all the input features properly, the statistical property of all the features were sent to the fusion server and a consistent method for scaling was obtained. Checking for quality of data required computing cross-site confusion matrices, and a policy based estimate of the quality of data could be done.

If the sites are providing their models at different times, the potential mismatch that can happen with sites that will join in the future is not known. As a result, any reconciliation among the formats, normalization approach, or quality can be done only among the fusion clients that are present at the start of the fusion process. That leads to the creation of an initial common model which expects input to be scaled according to some conventions, and some specific input format to be followed. Any model that is provided by a fusion client that joins after the initial common model needs to adhere to the norms that have been decided by the first set of participating fusion clients.

Predetermining the scaling function works well for numeric feature but, for categorical features, it is possible that new categories may show up when new clients join the fusion process in the future. A consistent encoding as binary values can provide a limited safeguard for the new values that may arrive in the future. The situation that can arise with encoding of categorical variables as new members join is shown in Table 7.1.

In the Table, it is assumed that initially only two members have joined the consortium, and for a specific categorical feature, the two members have the values as shown in the table. Initially, there are four categorical values among the two different participants. They could, using the algorithms outlined in Section 4.1.2, agree upon a set of encoding of the values shown in the top right sub-table. The bottom row of tables shows the new set of values when a new fusion client joins the consortium. This fusion client has a new value, Green, but it can not be encoded into the existing set of values which only allowed for four possible values for the feature, and is encoded as a set of all zeros. Another new value of Pink, provided by another new member is encoded as a set of all zeros.

Any value which is not in the original set of values will be encoded as all zeros, and the system can continue to function without warnings. However, the distinction among the new values can no longer be maintained. Both Green and Pink are encoded the same way in this approach. This may not be a desirable situation if the discrimination between the new values were important too.

While the One-Hot encoding provides a way to create a catch-all category which can capture new items that are now present in the data that makes up the original information to handle the situation. More than one such category value can be

Table 7.1: Categorical encoding issues with new fusion clients.

Consortium with Initial Members	
Fusion Client	Distinct Values
Client A	Blue, Red, Gold
Client B	Red, Blue, Yellow

Consortium with Initial Members	
Value	Encoding
Blue	[1 0 0 0]
Red	[0 1 0 0]
Gold	[0 0 1 0]
Yellow	[0 0 0 1]

Consortium with New Member	
Fusion Client	Distinct Values
Client A	Blue, Red, Gold
Client B	Red, Blue, Yellow
Client C	Gold, Green, Yellow
Client D	Gold, Pink, Red

Consortium with New Members	
Value	Encoding
Blue	[1 0 0 0]
Red	[0 1 0 0]
Gold	[0 0 1 0]
Yellow	[0 0 0 1]
Green	[0 0 0 0]
Pink	[0 0 0 0]

Table 7.2: Categorical encoding issues with new fusion clients.

Consortium with Initial Members			Consortium with Initial Members	
Fusion Client	Distinct Values		Value	Encoding
Client A	Blue, Red, Gold		Blue	[1 0 0 0 0 0 0 0]
Client B	Red, Blue, Yellow		Red	[0 1 0 0 0 0 0 0]
			Gold	[0 0 1 0 0 0 0 0]
			Yellow	[0 0 0 1 0 0 0 0]

Consortium with New Member			Consortium with New Members	
Fusion Client	Distinct Values		Value	Encoding
Client A	Blue, Red, Gold		Blue	[1 0 0 0 0 0 0 0]
Client B	Red, Blue, Yellow		Red	[0 1 0 0 0 0 0 0]
Client C	Gold, Green, Yellow		Gold	[0 0 1 0 0 0 0 0]
Client D	Gold, Pink, Red		Yellow	[0 0 0 1 0 0 0 0]
			Green	[0 0 0 0 1 0 0 0]
			Pink	[0 0 0 0 0 1 0 0]

kept reserved to expand in the future, since the original data never had those values, and one could assign them to the new values that were available.

An example for incremental encoding that allow for 4 new unknown values to arise in the future is shown in Table 7.2. This allows for 2 new values to be encoded, leaving room for 2 more values in the future.

The expansion of the encoding to have more values comes at the cost of additional computation load and needs to be balanced against the need to support anticipated new values.

When checking for value consistency, the cross site confusion matrix concept can still be used, except that the classes and labels of the new site data is compared to that of the existing model, and labels already existing need to be used. One challenge would be that if the new site provides a different label, it can only be captured as one new unseen class. The initial model would not predict any output for the unseen class. However, new sites may have data for the new class. In order to handle this situation, the initial model needs to make provisions for the original model to predict an unseen class.

The standard classifiers can not easily handle the concept of an unseen class or a new class [107, 108]. Furthermore, unlike the case where one needs to detect if the input one is seeing is a new class, the challenge for initial fusion is that some points corresponding to the new class ought to be generated. One approach to do that would be to ask each of the original sites to train N binary models for each of the N original class labels. Each of these models predicts where an input belongs to a class or not. The approach requires having some points that do not belong to any of the classes. These points can be identified in original data set because each of the N binary models will predict it as not belonging to its class. If no such points are present, some points which are not in any of the classes would need to be constructed. The augmentation of original data with these new points provides the ability for the original model to be ready to predict a new class.

A policy-driven approach for determining data quality can be used without any issues in asynchronous environments.

7.3 Ensemble-based Approaches

One approach to address the problems of synchronization is to use an ensemble of models, instead of trying to combine and fuse all the models together. The concept of a policy-based ensemble was discussed in Section 5.3. Using an ensemble-based approach for federated learning, each of the fusion model provides its independently trained model to the fusion server. The fusion server would assemble them into an ensemble which would be used to make predictions as an aggregate.

The ensemble that would be created would use the approach described in Section 5.3. However, the number of ensembles that would be used would vary over time. For the specific case of fusion sites joining and leaving the consortium shown in Figure 7.3, the ensemble would evolve as shown in Figure 7.4. The model provided by each site would be added to the ensemble as the different sites join the consortium. When a site leaves the consortium and would like to get their model and data removed, its model can be deleted from the ensemble. This gives an ability to easily deal with the membership changes that happen over time.

Ensembles have a significant advantage over approaches that fuse or combine the models together in that different sites need not coordinate and select a common model architecture. As long as the input values to each model are scaled consistently, and they are predicting the same output, i.e. each model in the ensemble

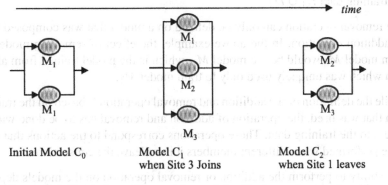

time

Initial Model C_0 Model C_1 Model C_2
when Site 3 Joins   when Site 1 leaves

Figure 7.4: Evolution of ensemble model with dynamic membership.

is estimating the same function, the model components can have any internal architecture. One of the sites could be providing a decision tree, one of the sites could be providing a neural network with ten layers of neurons, and yet another site could be providing a neural network with twenty layers. This flexibility means that the models could be trained completely independently, and do not need any coordination on model architecture at all.

Ensemble approaches would work well but have an issue that the size of ensemble would grow as more sites join the ensemble. For sites where information is collected from a handful of sites being used, an ensemble would be quite fine. However, if there are several sites that are contributing models for fusion, the ensemble can become very large and unwieldy. If there is a limit on how big the model ought to be when used for inference, an ensemble with several components could be problematic.

When combined with an approach to generate policies for selecting different models in the ensemble, as described in Section 5.3, this can yield a significant improvement in the performance of the ensembles [74, 75].

7.4   Conversion to Rule-based Models

The basic requirement to perform federated learning is an asynchronous manner is having the ability to combine pre-trained AI models when the data used to train the AI model is not accessible. If we have the ability to define an addition and a removal operation on models, then asynchronous training can be supported.

The addition operator would take two trained models and produce a combined model that captures the sum of all the patterns included in both the models. The semantic of addition operation can be defined on the basis of the training data that was used to create the model. Suppose model M_1 was created by training on data set D_1 and model M_2 was created by training on data set D_2. The addition of the two models $M_1 + M_2$ is the model M_3 that would be the model created using the training data $D_1 \cup D_2$.

The removal operation can only be defined on a model that was composed from an addition operation. In the above example, the effect of removing model M_2 from model M_3 would be the model M_1, which is the model trained from all the data which was uniquely used only to train model M_2.

While the definition of the addition and removal operation is based on the training data that was used, the operation of addition and removal has to be done without access to the training data. These operations correspond to the actions that need to be performed when different members join or leave the consortium.

The ability to perform the addition or removal operation on the models depends on the nature of the model. For neural network models, the addition and removal operations are hard to perform, and it is not clear how to do that in an asynchronous mode without any access to the original training data. However, for other types of models, e.g. when the model is a set of lookup rules, the addition and removal operations are fairly straight-forward to perform.

One approach to perform federation in an asynchronous manner for any type of model is to convert the model into an equivalent model which is based on a set of rules. The set of rules can then be subjected to addition and removal operations relatively easily. For the environment of coalition operations, where federated learning needs to be performed, this approach has been used to convert neural networks to equivalent rules [109]. The concept of attention [110, 111] is the fundamental basis for defining the interpretable rules. For any output predicted by a neural network, the attention mechanism identifies which part of the input space was most relevant to making the prediction. The interpretable rule mechanism defines the mapping from the attention to the prediction as a set of rules that provide a model which is equivalent to the original neural network.

Other AI models can also be converted to equivalent rule sets, e.g. decision trees can be converted to rules [112] and the rules can then be combined together to create a mechanism where multiple decision trees learned independently can be combined together [113].

In order to perform the addition operation on the rule sets, the fusion server needs to keep the different rule sets that are provided together. Each rule set predicts a certain output under a combination of conditions. The rules can be concatanated, and the conflicts among them, when different rules are predicting different outputs identified, e.g. by examining an overlap among the feature space defined by them [42, 114]. Such conflicts can then be resolved using human input or automated rules for prioritizing among the different sets that are provided by different parties. The end result is a set of rules that is the equivalent of the models provided as input.

We illustrate the process of combining different rule sets using a derivative of the algorithms detailed in [114]. For a training data set which consists of some n input features $x_1, x_2, \ldots x_n$ and an output y, each rule identifies some area in a feature space which is defined by having each dimension marked by one of the input features. For the sake of simplification, we will assume that the different features are completely independent of each other. While this assumption is not true in general, there are many approaches, such as principal component analysis [13] and multiple correspondence analysis [14], which can be used to convert dependent features into a set of independent features. Therefore, for explanation purposes, the independence assumption can be made without loss of generality.

Each rule effectively defines a set of conditions consisting of a set of combination of input features, and the corresponding output y corresponding to that combination. This can be viewed as defining a geometric region in the feature space, and any pair of rules that are considered may either overlap in their feature spaces, or be completely non-overlapping in the feature spaces. If there is no overlap, they do not conflict with each other. If there is an overlap, and the rules predict the same output, then they do not conflict with each other. However, if there is an overlap and the rules predict a different output, then there is a potential conflict.

The situation is illustrated in Figure 7.5. Four rules R_1 through R_4 are shown and they cover different regions in the feature space defined by two input features. Two input features make for easy illustration in a plane although the concept can be easily generalized to many dimensions. In the figure, rules R_1 and R_2 overlap, as do the rules R_3 and R_4. However, the overlap between rules R_3 and R_4 does not cause a conflict since both rules predict the output to be the same value (namely B). On the other hand, the region that overlaps between rules R_1 and R_2 has a conflict since one rule predicts the output in that region to be A while the other predicts the output to be C.

Examination of the regions of overlaps among the feature space can identify potential conflict among rules, specially rules that are obtained from different fusion clients. These conflicts can then be eliminated by defining some kind of prioritization guidelines as to which rule to select in the case of conflict. The existing algorithms for learning rules would usually result in a set of rules that are free of conflicts. However, when rules are collected from different clients, some conflicting situations may arise.

The conflict resolution can be based on the number of different fusion clients which are giving the same prediction, or some other criteria taking into account the reputation of the client providing the rules, or even using manual expert input. The ease of conflict resolution among rule sets provided by different fusion clients means that the task of adding models is relatively simple. The fusion server can take all the rules provided by the different fusion clients, analyze them to find the conflicts, resolve the conflicts, and obtain a new set of consolidated rules.

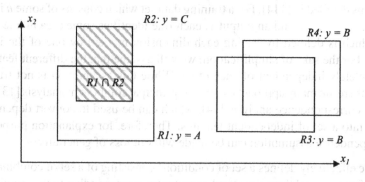

Figure 7.5: Illustration of rule conflicts.

Removing the contribution provided by a party whose model has to be removed is also straight-forward if the fusion server is keeping track of all the models provided by each participant. When a participant leaves, the model from that contributor can be removed and the remaining models converted to equivalent rule sets and combined.

A conversion to a rule-based model has another advantage. Rule sets can be inspected manually to determine if any over-fitting to the training data has happened. AI models sometimes learn patterns that may only be a construct of the data that was used to train them. Some of that over-fitting can be identified by manual inspection of the rules, but determining that over-fitting in other types of AI models, e.g. neural networks or decision trees may not be easy.

7.5 Data Generator-based Model Fusion

An AI model learns the patterns present within the training data, which we have generally expressed as the learning of a function. Once the function is learned, it can be used in many ways. While the primary focus of this book has been on the use of the function in the actual business process, the learned function can also be used for other tasks, and one of these tasks is generating new data that conforms to the learned function.

The basic idea in data generator-based model fusion is to learn a generator model over the data that is present at any site. The generator model can be used to regenerate the data points at the fusion server site. Once the generator models from all the data points have been used to recreate a representation of the data available at all of the sites, the system can train a new model that is based on the data that is representative of all of the data at different sites.

A simple example is shown in Figure 7.6. A fusion client has data set with several data points with one single input feature x and the output y, and the location of

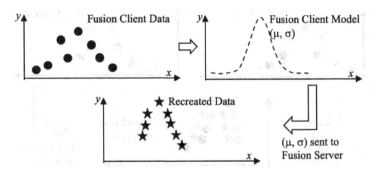

Figure 7.6: Example of a generator model.

data points is shown in the figure on the top left of the figure. From the domain, the fusion client knows that the relationship between the input can be modeled as a Normal distribution which is characterized by two parameters, the mean μ and the variance σ. The fusion client will calculate these two values and send it over to the fusion server. The fusion server would generate a number of data points taken from a Normal distribution and recreate the data locally. While the data is not exactly the same as that of the fusion client, it is close enough to be used as a proxy for that data. Characterizing the data as a probability distribution creates a generator model since the probability distribution can be used to generate the points on the data.

If there are several fusion clients and they all send a generator model to the fusion server, the fusion server can invoke each of the generator models to recreate data that is similar to that of the fusion clients. This completely regenerated data can then be used to perform any function, including the task of training an AI model. This process for recreating the data for federated learning from data at two different sites is shown in Figure 7.7.

While the example is shown using the generator model for only a simple distribution, there are many approaches to create a generator model. Similar approaches have been used to compute distributed clustering [115] and to compute distributed principal components of data distributed across several locations [40]. One example of generator models that can be used to recreate complex data is the technology for core sets [116, 117], which is effective for training machine learning models.

The recreation of data at different sites has many advantages for federated learning. The generator models are created and sent from each data fusion client to the fusion server, which means there is no need for coordination in how data is scaled or normalized across different fusion sites. The data generator models act as an efficient mechanism to recreate the data as needed, and a new model can be

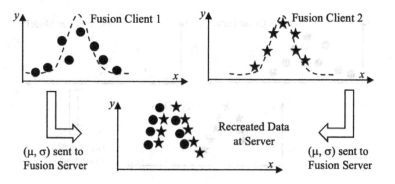

Figure 7.7: Data reconstruction using a generator model.

trained relatively easily when an existing member withdraws its data, or when new clients join. There is no need for all clients to agree to a common model architecture. If the data has been recreated, then each fusion client can be given a trained model using an architecture requested or desirable for the fusion client.

The effectiveness of the approach is limited by the fidelity of the generator model and its ability to recreate the patterns in the original data. Since the original data is not the same as the regenerated data, it is possible that some patterns may be missed out while new spurious patterns may be recreated. Despite these limitations, the ability of asynchronous training with minimum coordination provided by use of generator models means that this approach might be one of the most effective means for federated learning in real business environments.

7.6 Summary

There are many business situations in which the set of fusion clients sharing their models and combining them changes over time. In this chapter, we looked at some of the approaches that can be used to combine the models from new fusion clients without requiring them to provide their data. In these situations, the scaling of parameters that is established originally needs to be used for the future clients. Categorical values need to be encoded with an eye towards the future.

Ensemble techniques, using rule-based systems, and data generator-based models provide techniques to create models that can be trained in an asynchronous manner using federated learning techniques.

Chapter 8

Addressing Vertical Partitioning Issues in Federated Learning

The concept of data partitioning was introduced in Chapter 5. One specific case of data partitioning that is tricky to handle is that of vertical partitioning. Vertical partitioning occurs when the table representing the data does not have all of the columns at every fusion client. Looking at the data that was used as an example in Table 5.1 in Chapter 5, we can create an example of what may happen with data in that table in case of vertical partitioning.

In vertical partitioning, each record may be present at all of the sites, but not all features are present at all of the sites. An example of vertical partitioning is shown in Table 8.1. In vertical partitioning, each of the records is present at each of the sites, but each site is missing some of the columns. This vertical partitioning of data is problematic since the input to each model is different at each site. As a result, the models at each site are based on different inputs, and can not be combined together.

In practice, not all data records will be present at all of the sites, and real data will be partitioned as a mix of horizontal partitioning and vertical partitioning. As a result, practical solutions for federated AI in business contexts would need to use a combination of technologies described in Chapter 5 and this chapter.

Table 8.1: Vertical partitioning example.

Site A					Site B			
Index	F1	F2	Output		Index	F2	F3	Output
1	A	X	L0		1	X	0.3	L0
2	B	Y	L1		2	Y	0.2	L1
3	C	Z	L1		3	Z	0.3	L1
4	A	X	L2		4	X	0.4	L2
5	B	X	L0		5	X	0.8	L0
6	B	Y	L2		6	Y	0.9	L2

Site C			
Index	F1	F2	Output
1	A	X	L0
2	B	Y	L1
3	C	Z	L1
4	A	X	L2
5	B	X	L0
6	B	Y	L2

8.1 General Approaches for Handling Vertical Partitioning

When data is partitioned vertically, the set of features that is available at each of the sites is different. A representative scenario is shown in Figure 8.1. Three sites are shown, each of which has featured training data consisting of two features and an output prediction. The output prediction y is the same across all three sites. However, the set of features is different at each of the sites. When training independently, each of the three sites would be learning a different function, one predicting the output using two different inputs.

In the example shown in the figure, the first site has training data that will let it predict the output y as a function of two input features x_1 and x_2, the first site

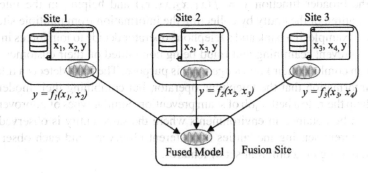

Figure 8.1: Example of vertical partitioning.

has training data that will let it predict the output y as a function of two input features x_2 and x_3 and site 3 has training data that will let it predict the output y as a function of two input features x_3 and x_4. Each of the three sites is learning a different function, which means that the algorithms mentioned in Chapters 3 can not be used in the manner discussed thus far.

The business needs to decide whether it makes sense to combine these three different function in the context of its operations. There are different manners in which business may want to use the result that comes from combining data from all of the different locations:

1. Refine Local Model: As a result of combining the models at the central location, each of the sites can get a model which is better than what they could train locally. If the input parameters x_3 and x_4 provide additional information which can make the model used by the first site $y = f_1(x_1, x_2)$ better, it may be useful to try to extract the model from the other site. In these settings, the central site can learn the broader function $y = f(x_1, x_2, x_3, x_4)$, and then, depending on the assumptions to be made about the values x_3 and x_4 at site 1, can result in a different and better prediction. As an example, a bank which collects different information about banking transactions in different countries can combine different models of the many countries using techniques such as augmented feature mappers (described later in Section 8.4).

2. Change Local Model: If, as a result of combining the models, the fusion site discovers that the collection of one of the input values, e.g. x_3 is very significant in the prediction of the output y, each of the sites may choose to change their data collection procedures and policies so that they collect the input x_3 as well.

3. Combine Insights: If the same entity is observed with different attributes by two different sites, then the two sites can combine their insights to learn more about that entity. In these settings, the central site is learning the broader function $y = f(x_1, x_2, x_3, x_4)$ and helping in the inference required of the entity by collecting the information from multiple sites. As an example, if a bank and a telephone operator decide to join forces in order to prevent scamming and fraud being committed on their customers, they can combine their knowledge for this purpose. The bank detects a different information that the telephone operator, but combining their models can help them do a better job of scam prevention. Similar types of improvements can be obtained in environments where the same entity is observed with different sensing modalities by different observers, and each observer is reporting on a different set of features.

When the local model is being refined, the usage of the AI model in the inference stage is the invocation of the function based on the features available at each of the sites. The broader function that is learned needs to be customized so that each local site can use it using only the features it has present locally. When the insights are being combined, the broader function is learned but there may be issues with sharing individual features of an entity during the inference phase. The exact scenario determines the limitations that are imposed on the type of information that each entity can share, and we will explore the scenarios further in Chapter 9.

If federated learning is being done in a context where there are sufficiently large number of data generation sites (e.g. 10 or more), it is likely that many of the sites share the same set of features. One can then identify groups of data generation sites which are using the same input features. This allows the usual federated model building to occur only within an individual group. However, there may be some situations when such groups may not be discernible, or there may be several groups with only one member site. In those cases, partitioning all sites into separate clusters where federated learning can be done may not be viable, and the other schemes described in this chapter need to be used.

In the rest of this chapter, we look at the different manners in which the situation of vertical partitioning can be handled.

8.2 Rule-based Approach

Decision rules form a well-established field of traditional machine learning which have been around since the 1970s. The biggest advantage of decision rules is that they are easy to understand, explain and combine. This ability allows them to be used for asynchronous federated learning, as discussed in Section 7.4 of Chapter 7.

When the models provided by different clients use different input, the rules would use different sets of inputs to make their predictions. For the specific example shown in Figure 8.1, the rule sets from site 1 would consist of a variety of rules where the conditions are defined by a combination of two input features x_1 and x_2 and are agnostic about the value s of x_3 or x_4. Similarly, site 2 would have rule sets dependent on the vales of f x_2 and x_3 and ignoring any value of x_1 and x_4, whereas site 3 has rule sets as a combination of two input features x_3 and x_4 and is independent of any values defined by x_1 or x_2. The projections of the rule sets from different sites on the relevant two dimensions are illustrated in Figure 8.2. The rules provided by any individual site would be free of internal conflicts, so the areas mapped out by each of the rules do not overlap with each other.

Since there are four input features, the maximum and minimum value of any of the input features by any site can be computed. The ranges for x_1 will be computed only by the rules provided by site 1, whereas the bounds for x_2 will be provided

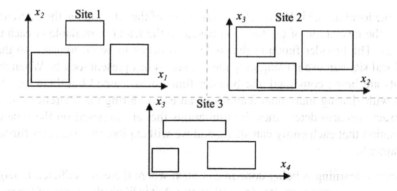

Figure 8.2: Rule sets with vertical partitioning.

by both sites 1 and 2. Similarly, bounds for x_3 will be provided by both sites 2 and 3, and the bounds for x_4 are only provided by site 3.

The rule sets provided by each of the sites can now be expanded to cover all of the four input features, assuming that the values are valid for all the values of the features that are missing. As an example, each rule from site 1, which consisted of only two specified features, is assumed to be valid for all values of x_3 and x_4, with analogous extensions for the other sites. As a result, we have a rule-set which is defined for all four input features from each of the sites.

This rule-set can then be combined and any resulting conflicts among the different rules and their overlapping areas can be identified, and resolved. The net result would be a set of consistent rules which are defined in terms of all the four input features. For each site, one can take the projections of the resulting rule-set only onto the features available at the site, and the associated set of rules (once again after resolving any conflicts) provided back to each of the site as an augmentation to the model that they had provided. If insights need to be combined, the global model with all of the input features can be used in a centralized manner or in a distributed model to provide this combination.

An alternative approach would be to do a joint training of the rules across all of the data sets using a distributed association rule mining algorithm [118]. This results in a set of rules that is the combination of all of the features that are available at any site. Each site could then use the subset which only uses the features available locally.

8.3 Feature Prediction Approach

The main challenge with using the rules to combine insights from different sites is that they are assuming that different features are independent and, thus, extending any missing features in a rule to be valid for all combinations of features is valid. While techniques to convert features into independent components [13, 14] are known, there may be several data sets where such a conversion may not be inadequate to capture the data patterns properly. In those cases, a scheme to be able to predict the missing features at every site may work well.

Considering the task of learning the global model as a function that is based on the union of all features that are available across all of the sites, we are use an iterative federated learning approach across all of the sites to build models for different features. This process consists of three steps:

1. Determine the order for feature prediction: This will result in a sequence in which features that are missing at any site ought to be predicted.

2. Use Naive Federated Learning to create feature predictor models: This results in a set of models that each site can use to augment their data to consist of all the features.

3. Extend and Federate: Learn a federated model to predict the output across all the feature at all of the data.

In order for this approach to work, at least one feature should be common across all of the sites. The feature which is missing from the smallest number of sites can be selected to be the first one to be predicted. Then a federated model which can be used to predict the missing feature from the common features across all of the sites is trained. This federated model can now be used to predict this feature so that the number of common features across all of the sites is incremented. Then the next feature to be predicted can be selected. Once all of the features have been predicted, a federated learning mechanism to predict the output from all of the features can be learned. Of course, other order of feature prediction, e.g. based on the relative size of missing features, or any other metric, can also be used in this process.

As an example, suppose there are four sites with the following configurations:

- Site A has features x_1, x_2, x_3, x_4 and output y
- Site B has features x_1, x_3, x_4, x_5 and output y
- Site C has features x_1, x_2, x_4, x_5 and output y
- Site D has features x_1, x_4, x_5 and output y

In this particular case, features x_1 and x_4 are common across all the four sites. Feature x_2 is missing from two sites, feature x_3 is missing from two sites and feature x_5 is missing from one site.

One possible order for missing feature generation would be to predict feature x_5 first followed by prediction of feature x_2 and finally the prediction of feature x_3.

In order to make this prediction, all the four sites would initially create a joint model M_5 which would take the input of x_1 and x_4 and predict the output x_5. In other words, the joint model will be created by using a Naive federated learning algorithm to learn the function f_5 such that $x_5 = f_5(x_1, x_4)$. Model M_5 can be trained by using federated learning between the sites B, C and D. The model is then used by site A to use the inputs x_1 and x_2 to generate the predicted values for x_5 at A.

As a result of this generation, the four sites now have the following configurations:

■ Site A has features x_1, x_2, x_3, x_4, x_5 (generated using M_5) and output y

■ Site B has features x_1, x_3, x_4, x_5 and output y

■ Site C has features x_1, x_2, x_4, x_5 and output y

■ Site D has features x_1, x_4, x_5 and output y

Now, features features x_1, x_4 and x_5 are common across all the four sites, with features x_3 and x_3 missing from two sites each. A process of federated learning can now be repeated to learn the function f_2 which can predict x_2 as a function $x_2 = f_2(x_1, x_4, x_5)$. This model M_2 can be trained using federated learning between sites A and C. The resulting model M_2 is shared with sites B and D, and they can now generate their esitmates for the feature x_2 locally using the model M_2.

As a result of this generation, the four sites now have the following configurations:

■ Site A has features x_1, x_2, x_3, x_4, x_5 (generated using M_5) and output y

■ Site B has features x_1, x_2 (generated using M_2), x_3, x_4, x_5 and output y

■ Site C has features x_1, x_2, x_4, x_5 and output y

■ Site D has features x_1, x_2 (generated using M_2), x_4, x_5 and output y

Now, features features x_1, x_2, x_4 and x_5 are common across all the four sites, with feature x_3 missing from sites C and D. A process of federated learning can now be repeated to learn the function f_3 which can predict x_3 as a function $x_3 = f_2(x_1, x_2, x_4, x_5)$. This model M_3 can be trained using federated learning between sites A and B. The resulting model M_3 is shared with sites C and D, and they can now generate their estimates for the feature x_3 locally using the model M_3.

As a result of this generation, the four sites now have the following configurations:

■ Site A has features x_1, x_2, x_3, x_4, x_5 (generated using M_5) and output y

■ Site B has features x_1, x_2 (generated using M_2), x_3, x_4, x_5 and output y

■ Site C has features x_1, x_2, x_3 (generated using M_3), x_4, x_5 and output y

■ Site D has features x_1, x_2 (generated using M_2), x_3 (generated using M_3, x_4, x_5 and output y

Now, each of the sites have all features, and a final round of federated learning can be used to learn a model M_o which can predict the output y using the inputs. This final model is available to all of the sites.

When the different sites have to make a prediction, they would each have to operate in a slightly different manner as follows:

■ Site A will start with the values of x_1, x_2, x_3, x_4, use model M_5 to predict x_5 and use model M_o to predict the output y

■ Site B will start with the values of x_1, x_3, x_4,x_5, use model M_2 to predict x_2 and use model M_o to predict the output y

■ Site C will start with the values of x_1, x_2, x_4, x_5, use model M_3 to predict x_3 and use model M_o to predict the output y

■ Site D will start with the values of x_1, x_4, x_5, use model M_2 to predict x_2, use model M_3 to predict x_3, and finally use model M_o to predict the output y

Instead of using a single model for prediction, each of the sites now has a different approach for prediction, which is based on concatenating different models together. The resulting approach is shown in Figure 8.3. Each of the sites has been using Naive federated learning to help each-other to train intermediary models to predict the features, and once all the features have been predicted, the predicted values are used to estimate the output.

The efficacy of the approach depends upon the ability to predict each of the missing features correctly from the features that are available at the site. This would depend on the data and the domain of the application.

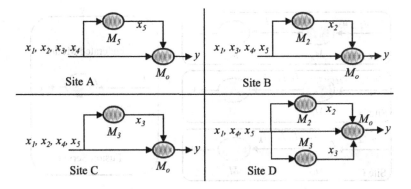

Figure 8.3: Example with feature prediction approach.

Other variants with different approaches for filling in the missing features can also be used, such as using a randomized selection of missing features at each site [119] or using semi-supervised learning algorithms to predict missing features [120].

8.4 Feature Mapper Augmentation

The concept of using an AI model to predict the missing features and use several models to generalize inputs to a common model can be viewed as a special case of model harmonization, i.e. while each site is learning different aspects to solve the same problem, and they can benefit from sharing their insights with each other. The common model would be expressed in terms of a latent space, i.e. in terms of a set of features that are hidden or latent, and provide the desired output. The set of different features available at each of the sites may be different but it can be mapped onto the latent space.

Expressing in terms of the function estimation model, we assume that there are a set of K latent features $l_1, l_2 \ldots l_K$ and a common function is being learned at each of the sites, namely $y = f(l_1, l_2 \ldots l_K)$. Each site also has a different set of features, e.g. site A may have a set of N features $u_1, u_2, \ldots u_N$, site B may have a set of M features $v_1, v_2, \ldots v_M$, site C may have a set of P features $x_1, x_2, \ldots x_P$ and so on. Each site would have a local AI model that maps its input features to the set of latent features, and there would be a common AI model which maps the latent features to the common output y. The challenge is for each site to train the models so that the core model is trained in a federated manner while each of the individual sites is training its own feature mapper in an independent manner. However, each site only has data that predicts its output y as a function of the features it has collected locally, so the latent parameters are not known.

The situation is shown in Figure 8.4. Three sites are shown with their input features and the approach that they want to follow of having a feature mapped

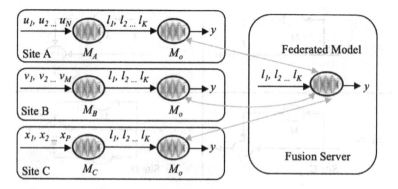

Figure 8.4: Example with feature mapper approach.

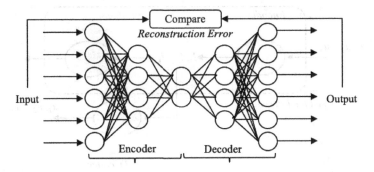

Figure 8.5: Example of an auto encoder.

followed by a shared model to predict the output. The shared model M_o is the only part which can be federated since it uses a common set of inputs and outputs. The feature mappers M_A, M_B and M_C are very specific to the site that has the local data, and have to be trained independently.

One of the ways in which this problem can be addressed is by coupling a neural network which can extract latent space features with a neural network that performs the prediction using the latent space features. A common method for extracting features is the use of auto-encoders. An auto-encoder is a neural network that consists of two parts, an encoder part that maps the original input to a set of latent space features, and a decoder that maps the latent space features back to the original. The comparison of the reconstructed input can be used as a measure of goodness to determine when the auto-encoder has been trained completely. A sample auto-encoder is shown in Figure 8.5.

A auto-encoder can then be coupled with a neural network to train a network that performs the or prediction of the output y. Two parameters, the number of latent features that will be used, and the relative weight between the accuracy of the feature extraction process and the accuracy of the predictor, are determined in advance. The system can now be trained by each of the members in a manner where the predictor is trained using the federated approach, while the auto-encoder is trained locally, with the learning process jointly optimizing both sections of the neural network.

Once the neural network is trained, only the encoder part is used with the predictor during the inference process for each of the local sites. This provides each of them with a model that works with their input. The task of mapping features to a common representation is learned by the auto-encoder.

The model from the perspective of any single site is shown in Figure 8.6. The three neural networks, the encoder, the decoder and the predictor are combined in the manner as shown in the figure. During the training phase, the output of

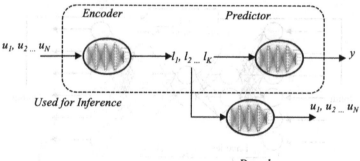

Figure 8.6: Models at each site for learning feature mappers.

the predictor and the decoder are used to determine the weights of the neurons in the neural network. During the inference phase, only the two networks, the encoder and the predictor, are used, and are shown in the dashed rectangle within the figure.

8.5   Federated Inference

Federation and collaboration among different sites need not be done only during the model learning stage, but can also be used during the inference stage. After all, the reason for training and creating an AI model is to use it during the inference stage. When the input features are partitioned, a use of federation during the inference stage may be more effective than trying to create a shared model.

A simple (although perhaps a little futuristic at the time of writing this book) use-case of federated inference would be when a self-driving car approaches an intersection. The sign at the intersection may be partially occluded, e.g. from an overgrown tree branch, or because the sign may have been bent in a way that the car cameras do not get a good angle on the sign. If there were other nearby cars with a better view of the object, the cars could communicate among each other to exchange their perspective on what the sign could be, and each car can be more confident in their decision making.

When the same entity is being observed by two different sites which are collecting different types of information about the entity, a common model can not be built across both sites because the inputs are different. However, if the two sites are collaborating together, they may still be able to share the output from their models about the entity during the inference stage, and sharing inference results would help each site in making their decision. This is the use of Federation during the inference stage as opposed to the model building stage.

In a more realistic current business scenario, criminals target innocent civilians to try to get access to their money by making telephone calls and sending spam emails to them. While the eventual fraud would happen with a victim transferring money from their accounts, which involves a bank or another financial firm, the procedure also involves the criminal interacting with a telecommunications company or an Internet Services Provider. Since the same entity is observed by multiple enterprises, they could share information about the suspected criminal with each other to do a better task at fraud prevention. While normal sharing of private information among service providers like banks and telephone service providers may be prohibited in general under privacy regulations like GDPR [121], it is allowed for narrow targeted objectives such as fraud prevention.

Another practical use-case is that of multi-modal sensor information fusion. When the same object is observed through multiple different types of sensors, each sensor provides an indication of what it is seeing. This can be used to identify the type of object being seen through different modalities, e.g. use both sounds and vision to identify an animal observed in the wild-life, or use both sound and vision to determine if the machines on a factory floor are operating normally, or experiencing some abnormal situation. The AI models to be used for each modality of sensing would be different, but their results can be federated together.

Federated inference can be viewed as an implementation of a distributed ensemble of AI models [48] where each of the models in the ensemble comes fully trained. The results from different ensemble models need to be combined together. The combination can be done using a majority voting, or using a variation of policy-based techniques, as described in Section 5.3 in Chapter 5.

The process of federated inference among different sites will usually happen in the manner shown in Figure 8.7. Multiple sites observe an object and they each extract a different set of features from the object. In the specific example that is shown, features $u_1, u_2, \ldots u_N$ are extracted by site A, and features $v_1, v_2, \ldots v_M$ are extracted from site B. There may be other sites extracting their own features as well. Each of the sites has their own model which has been pre-trained to use the features that they extract. Site A has a decision of y_A while Site B has a decision of y_B. All of the sites exchange their decisions with each other using some exchange mechanism. The easiest exchange mechanism would be the use of a server to exchange the information, but there are several other mechanisms for such exchange that do not require a server. Once each site has information about all of the other decisions, it can decide on the basis of a majority vote or another resolution mechanism to determine which final decision to use.

When different models are used, meta-data about different models can be used to determine how much credence to provide to each of the inferences provided by the models. A weighting mechanism, which is borrowed from the concept of policy-based ensembles, could be to see the distance of the inference point from

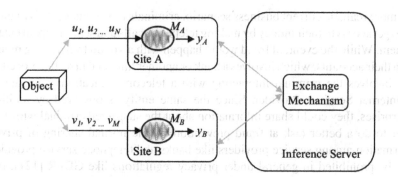

Figure 8.7: Process of federated inference.

the region in the feature space that the model has been trained upon. Each site will need to provide that information in addition to its inference. Other metrics that can be provided would include the confidence in their prediction by each model.

Federated inference provides the ability to combine insights from many different sites in order to make an improved decision.

8.6 Summary

Vertical Partitioning of data arises in many business contexts. This is a situation where different sites have collected information to predict a common function that will be used in their business context, but the input set of features used within the model is different at each site. In order to handle this situation, one can try to extend the features so that all sites can use a common model using feature prediction or a feature mapping mechanisms. One can try to learn rules across multiple features from different sites as well. Instead of trying to build a common model or enrich their local models, another viable option for many business operations would be to use the result of inference from many different sites, resulting in a federated inference.

Chapter 9

Use Cases

In Chapter 2, we discussed several industries that provided the motivation for developing federated AI. Within each of these industries, several use-cases of federated AI exist.

In this chapter, we bring together the various approaches described in the previous chapters to discuss how federated AI can be used in some real-world use-cases. The use-cases in this section are just examples that combine the approaches discussed previously in a manner to solve a business problem. Many other combinations can be made in order to address other business problems.

9.1   Collaborative Scam Detection

In this section, we look at the specific challenge of detecting criminals who may be trying to defraud members of the public. The specific use-case that we will target will be that of people who call upon the members of the public to try to get them to transfer money to the fraudsters. The fraudsters may use a variety of techniques for this purpose, encouraging the victim to transfer funds to them using mechanisms that are hard to track and reverse, e.g. purchase of gift cards. This use-case was explored at the request of a bank in Europe which was investigating approaches to reduce fraud among its customers.

In order to prevent this type of fraud, the proposal was to form a consortium of banks and telephone operators. While different banks compete with each other, they would also collaborate with each other for prevention of fraud which impacts all of them collectively. Similarly, telephone operators would be willing to collaborate in this venture since this would provide a safeguard against the misuse of their networks.

The objective was to identify fraudsters through a collaborative analysis of their behavior. One would expect a fraudster to have a different calling pattern than that of the normal members of the public. However, the telephone calling behavior of the fraudster may not be very different from that of marketing or survey companies that call upon a broad segment of the population to get their opinions, or to buy into any promotional activities around their products. If telecommunications companies were to try to flag fraudsters on the basis of their calling patterns alone, they would have very high false positive rates in such detection.

Similarly, the banks may notice that some of their clients are showing unusual behavior, such as receiving unexpected high levels of transfers to their accounts, initiating gift card purchases, or other activities that are outside the norms of other clients. These can be identified via AI-based anomaly detection algorithms (see Section 1.8.3 in Chapter 1). If the fraudsters are using their own accounts to receive the transfer, their deposit patterns will be anomalous compared to normal users and the banks may be able to detect them. If the fraudsters are not transferring funds directly into their account, but instead persuading their victims to send them funds using bank-issued cash cards/gift cards, the banks may detect unusual patterns in the issuance of those cash cards, or an unusual patterns in the spending made on cash cards they have issues. Even if the fraudster is using direct transfers to their own accounts, the transaction patterns of a fraudster may be similar to businesses which mail online products, and may be hard to distinguish from normal transnational pattern. It would be even harder to get the patterns identified from the use or issuance of cash cards.

While a fraudster may have patterns similar to that of some legitimate businesses in telephone-based calls, and the fraudster may have patterns similar to that of some legitimate businesses in their financial transactions, they are less likely to have the same patterns with legitimate businesses in both industries. If a consortium of banks and telephone companies can be brought together to collaborate around fraud detection, they are likely to perform better.

Let us explore how a solution to collaborate on the scam detection process can be developed.

9.1.1 *Collaboration within a Single Industry*

Let us consider an initial situation where only banks are going to be part of the consortium trying to prevent the fraudulent transactions. The discussions in this section apply equally well to the situation when only the telephone companies are collaborating together to prevent the fraudulent transactions. Since the collaborating companies are in the same industry, they are performing similar type of functions and would be maintaining similar information about their clients. Individually, each of the banks may have selected a different format or schema to store the information. In order to collaborate together, the banks would need to agree upon a common schema. They would translate their data into the common schema. With all the data in a the common schema, the banks can learn a shared model to detect and prevent fraudulent behavior using the federated learning algorithms described in this book.

In addition to agreeing on the common schema, the participants would need to agree on some other items, such as the frequency with which the model predicting the fraudulent behavior would be updated, and the mechanism for inference to detect fraudsters using the model. Once a shared model is learned, each bank may retrieve the shared model, and choose to run their transactions through the model locally. Alternatively, they may decide that they would like a consortium server to check on every fraudulent transaction. The latter is likely to face more resistance in the real-world, since the use of inference would expose more private information from a participating bank to the consortium servers. The former model, where the banks can share their models and not share any specific personal information, is more likely to be adopted. Nevertheless, depending on the nature of business agreement among the companies, either of these options would be viable.

For the sake of illustration, we will assume that the joint model is trained once every month, and then distributed to each of the banks. The banks would then use the model to check their own transactions, and provide a second layer of manual checks on any transactions that appear to be suspect. With the manual checks, it would be acceptable (though not desirable) if some normal transactions would be flagged as being fraudulent since the manual check would correct that mistake.

Once the banks have agreed to participate in the task of collaborative fraud detection, each of the participating banks would also need to set up the infrastructure required for collaborative fraud detection. One possible solution would be for the consortium to run a shared model fusion server in a cloud provider's system, which is used to collaboratively train the model. For model training, a possible infrastructure may look like the one shown in Figure 9.1. Each bank maintains its transaction data in a proprietary format in its green zone. It would translate the information to the common agreed upon format to its servers in the yellow zone within its premises once a month to enable checking of the models. The definition of yellow zone and green zone would be as described in Section 6.4. Yellow zone provides limited access to external systems, while green zone houses the systems only accessible within the bank. The green zone systems would convert the data in the common format into the model trained on local data and copy that to the yellow zone systems. A service setup in the public cloud is used to combine the different models provided by all of the banks into a combined model.

Before the model is trained, the bank systems in the yellow zone would need to coordinate with the fusion server in the cloud in order to determine the normalization conventions for their data, as described in Section 4.1.2 in Chapter 4. The

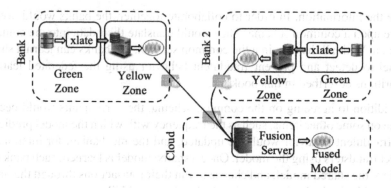

Figure 9.1: Infrastructure for collaborative model building.

bank systems in the yellow zone have the role of fusion clients, as per the structure shown in Figure 3.1. Assuming that the business agreements enable call-in from the fusion server to the fusion clients, the banks can operate in a synchronized manner to create the model using the algorithms described in Chapter 3. However, the banks may prefer to train the models in an asynchronous manner, using the techniques described in Chapter 7, if they want new banks and telephone companies to join the consortium in the future.

Once the model is trained, it is provided to each of the participating banks. The fusion clients can retrieve them to the yellow zone fusion servers, and systems in the green zone can retrieve them from the yellow zone. This model would be valid for use for about a month, at which point the next model training session would happen. During that month, each bank would be taking their transaction data, running it through the model to identify potential fraudulent behavior, and marking any suspicious transactions or suspicious clients for further manual investigation. This identification can be done in real-time on a transaction by transaction basis, or potential fraudulent customers can be identified by checking their transactions on a daily basis.

9.1.2 Collaboration Across Industries

While the banks and telephone operators may agree in principle that collaborating across their industries will lead to a better detection of fraudulent transactions, they have to overcome the challenge that they are in different lines of business and collect different types of information. While the banks can agree among themselves on a common schema for their transactions, and the telephone companies can agree among themselves on a common schema for their transactions, there is no common schema that works across the telephone companies and the bank companies. As a result, a common model across multiple industries can not be built.

Across both types of industries, there will be some common fields for all the clients, since the telephone companies and the banking companies are all serving the same population. The telephone numbers, email addresses and mailing addresses provided as contact information to the banking companies, are usually also provided to the telephone companies and are some of the common fields across both types of companies. Some telephone companies may also have the account number and the bank information for their clients to pay for their bills. These common fields provide a mechanism to match clients across both of these industries.

This commonality means that the techniques used for vertically partitioned data described in Section 8 can be used for collaboration across the different industries. As an example, we can use federated inference techniques (Section 8.5) as a methodology to combine the insights from the two different industries. Assuming the banks and telephone operators are identifying suspected fraudsters among their client set by analyzing the transactions on a daily basis. They could check on suspected fraudsters with other banks or telephone operators by sharing the common features, and collecting the inference results from other parties. This collected information would help them make a better determination.

To share the results from other participants, each of the consortium members may choose to share the identity of clients suspected to be fraudulent with the fusion server. This list of suspects can share the resulting list with each of the other members as requested. The consortium members would get the set of all the suspect fraudulent members from other members, and can compare that against their own client set in order to determine what the majority of the members are classifying their clients to be. Note that any member not specified explicitly as a potential fraudster by the other members can be assumed to be a normal user. This allows for sharing of inference results which is bandwidth efficient (only the small number of suspected fraudsters are in the information exchanged) and does not require any participant to do anything other than making outbound calls to the cloud site. Such information sharing for fraud detection and prevention is usually permissible under the data privacy rules, such as the European GDPR [121].

If there are multiple banks and multiple telephone companies, the banks can create a joint bank fraud detection model using federated learning and the telephone operators can create a joint telephone fraud detection model using federated learning. Note that the same model can flag the same user as fraudulent in one bank and not fraudulent in another bank since the two banks would have different set of features characterizing the user. The banks and telephone companies could collaborate with each other using federated inference.

9.1.3 *Effectiveness*

The effectiveness of any AI model in any use-case depends upon the patterns contained in the data used for training and testing the model. The results in any situation would depend on the data, the model learning algorithm used, the model architecture chosen, and the federated learning approach selected. Consequently, the results discussed in the section should not be considered as holding under all scenarios and all circumstances. However, they do provide representative data points for comparative benefits of different techniques and approaches.

The results in this section show the effectiveness of three different approaches for detecting fraudsters in a community. For this particular set of results, it is assumed that two banks and two telephone operators are part of a consortium, and they examine their transactions daily to identify any clients that may be potential fraudsters. Results are provided for three different operational modes. Synthetic transactions were generated to simulate behavior of a set of normal companies, normal consumers and fraudsters to train and test the models. The generation of synthetic data gave access to the ground truth regarding which of the clients of banks/telephone companies in the system were normal users and which were fraudsters.

The first approach is that of independent operation. In this particular mode, each bank and each telephone company operates independently. They do not share their data with each other. Each company operates its own system to identify fraudsters, which looks out for anomalous transactions conducted by its clients. In this analysis, the algorithm used is an unsupervised one which flags anomalies by means of an auto-encoder. The auto-encoder is run over the transactions of each company at the end of the day to determine anomalous transactions. The pie-chart at the top for each company shows the breakdown of how effective the company was is in characterizing their clients into the right category. Since each company has different combinations of features and shared model, the results could be different for different companies. The pie-chart at the bottom only considers the set of active fraudsters and shows whether the fraudsters are identified correctly or not.

For the pie-chart at the top, there are four possible outcomes (i) A fraudster is flagged correctly as a fraudster (ii) A fraudster is missed and considered a normal user (iii) A normal user is identified correctly as a normal user and (iv) A normal user is incorrectly flagged as a fraudster. Each of the four categories is shown using different shades of the slice in the pie-chart. Ideally, the system would consist only of categories (i) and (iii) but since no learning algorithm is perfect, some mis-identification can occur.

Given the existence of the second layer of manual checks, it is more important that all fraudsters are identified correctly and having some normal people incorrectly flagged as fraudsters is acceptable (but undesirable). The bottom pie-chart shows the outcome for the fraudsters, which is in two categories, whether they are flagged correctly or missed out upon. This pie-chart shows a simpler mechanisms about the effectiveness of the mechanism by using the fraction of fraudsters that are detected correctly as the performance metric.

When the companies do not collaborate with each other, the performance of any individual company in identifying fraudsters is not very good. A large fraction of the fraudsters are missed, and a substantial percentage of normal users are incorrectly flagged as fraudsters. While this may not impact their eventual ability to complete the transaction, it does waste the time of bank or telephone employees as they process them in the next step.

If only the banks and telecommunication companies are able to form their own respective consortia, they can build a shared model to track the fraudsters. The shared model would identify abnormality of usages across multiple companies.

In this particular case, both banks will be sharing their model with each other using the setup described in Section 9.1.1. As a result, both banks get to learn a model for fraudster identification which understands the transaction patterns across both of them. Similarly, each of the telephone companies gets to learn a model for fraudster identification capturing the patterns for telephone calling across all telephone companies. It would be expected that the resulting ability of the model to identify fraudsters would be better.

The results, shown in Figure 9.3, confirm that there is improvement over the case where each company was trying to operate on its own, as shown in Figure 9.2. The percentage of fraudsters identified correctly has increased, and the percentage of normal users who were incorrectly identified as fraudsters has decreased. As expected, sharing of information across all companies in an industry results in a better fraud detection scheme.

Figure 9.2: Results for independent detection.

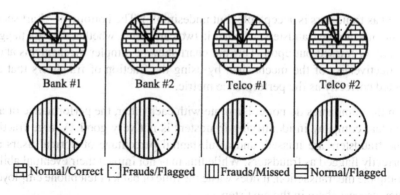

Figure 9.3: Results for detection by collaboration within industry.

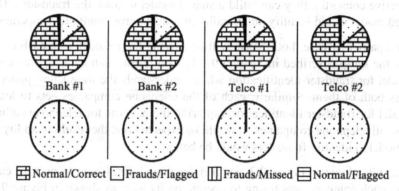

Figure 9.4: Results for detection by collaboration across industries.

If there is a cross-industry collaboration, banks and telephone companies can share information with each other using a combination of federated learning and federated inference, as described in Section 9.1.2. The results from cross-industry collaboration are shown in Figure 9.4. In this particular case, all the fraudsters are identified correctly, and there is a very small percentage of normal users who are flagged incorrectly as fraudsters. This performance is significantly better than in the other two approaches.

While one can not claim that equally good results would be obtained in all types of fraud detection, it would be fair enough to claim that federated learning across companies within the same industry would be better at detecting fraud that each company trying it on their own. Similarly, cross-industry collaboration using federated inference can provide significant improvement over collaboration within a single industry.

Real-world financial fraud comes in many different forms [122] and can be identified using many different AI/ML-based approaches. While a perfect identification may not always be possible, the results in this section provide affirmation of the expectation that federated learning and federated inference can improve the status quo.

9.2 Federated Network Management

Collaboration across multiple organizations in consortium settings can pose its own challenges in execution. Getting multiple organizations to agree to collaborate with each other is a non-trivial exercise. However, there are many applications of federated learning within a single company.

In this section, we will examine a use-case of federated learning required within the context of a single company, namely a telecommunications network operator. The specific problem we will look at is the management of a cellular network. The use-cases were developed for a network carrier in the United States.

A cellular network provides the infrastructure which allows modern mobile phone communication to happen seamlessly. The generic structure of a cellular network at a very high level is shown in Figure 9.5. More details of the network infrastructure can be found in reference [123].

In a cellular network, mobile phones connect to a cell tower over the air in the initial leg of data being exchanged with a server in the Internet. The different cell towers are interconnected via a mobile network operated by the cellular network operator. The mobile network runs the communication protocols (e.g. the 4G or 5G protocols) which enable the mobile phone and cell-tower equipment to interact with the systems providing access to the Internet. The cellular network operator also runs a service network which is used to support various support functions required for network operations, which include billing, management and operations support systems. One specific function in the service network is

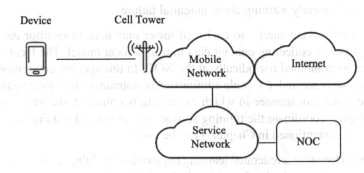

Figure 9.5: Structure of a cellular network.

the Network Operations Center (NOC), which houses various systems used for monitoring, trouble-shooting and correcting any issues that may arise during the operation of the cellular network.

One of the challenges in the management of the cellular network is its large scale. In a typical cellular network environment, there would be several tens of thousands of cellular towers, around a thousand network data centers acting as intermediary locations for protocol processing, and a few exchange locations where they interface with the Internet. The Network Operations Center needs to be monitoring the machines at several thousand locations, understand if any of them may be experiencing a problem, and then take corrective action. Given the extensive amount of instrumentation in network communication infrastructures, any failure can be detected and reported to the NOC fairly rapidly. Additionally, events that are happening at a cell tower can provide indication of failures that may happen in the future. These events include the entries recorded in the system logs of the equipment as well as messages exchanged among different machines with whom equipment at cell-towers communicate to run the network.

In order to build such a model, the NOC needs access to the events and failure information at various cell-towers. Failures are notified to the NOC, but with the events that are happening at the equipment at each of ten thousand cell towers (this number will be for a small cellular operator) and then analyzed over the course of a day to understand the correlation with failure indications, the NOC needs to collect around ten billion events every day. Just the collection of this data to build a predictive model for failures would use up a significant amount of bandwidth in the mobile network which the network operator may prefer to use for revenue-generating traffic from its mobile phone users.

Instead of an approach where the NOC collects data from all the cell-towers, it would be a better approach to let each cell-tower train its own local model that can learn the correlation between the events happening at the cell-tower and the likelihood of a failure in the future. This model can then be federated across different cell towers and used to constantly monitor the events at the cell tower in order to get an early warning about potential failures.

Failure events occur rarely, so each cell tower may have to monitor data for a few days before collecting enough data to train a local model. The local models can then be combined periodically at the NOC. In this specific environment, all cellular towers are under a single administrative domain, so the system can chose a single consistent manner in which event data is collected, the model trained, and can even coordinate the training so that a synchronized training algorithms like the ones mentioned in Chapter 3 can be used.

The reason for using a federated learning approach for training models in network management is the significant saving in bandwidth that can be obtained while creating the model. Using the equations for speedup of federated learning [124], and the fact that there are a large number of cell towers in the system, the relative gain in the time taken to train a model can be computed using two primary factors, first the reduction in model size and second the relative speed of the network measured by what we can refer to as the *network ratio*. The network ratio ranges from zero to infinity and compares the relative speed of training a model at the edge on a set of training data compared to the the time taken to transfer the data over the network to the NOC to train the data at a central location. If the network is really fast, this ratio would be close to zero. On the other hand, if the network is relatively slow, the network ratio would be much higher.

The metric of *model size reduction* measures reduction in model size as the ratio of model size to the size of data used to train the model. A model reduction of 0.1 means the model is one tenth the size of the training data. A model reduction of 0.01 means the model is one hundredth the size of the model training data.

Federated learning results in training an AI model much faster than trying to move the data to the NOC. The speed up in time taken to train the model is shown in Figure 9.6. A system with ten thousand cell towers is analyzed. As shown in the figure, the speedup is linear with the reduction in model size, and the slope of the speedup is determined by the network ratio.

To understand the impact of network ratio, we plot the speed up against the network ratio for two conditions, the first one where the network ratio is less than 1 (i.e. the network is relatively fast and can transfer data much faster than it takes to train the model), the second is where the network ratio is greater than 1 (i.e. the network is relatively slow and model training locally is faster than the time it takes to transfer the data).

Figure 9.7 shows the speed up when the network is relatively fast compared to the task of training a model. When the network is really fast, federated learning

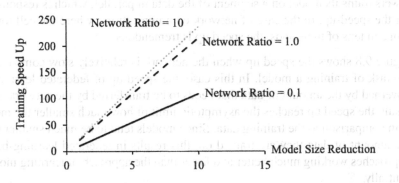

Figure 9.6: Relative performance of federated learning to centralized approach.

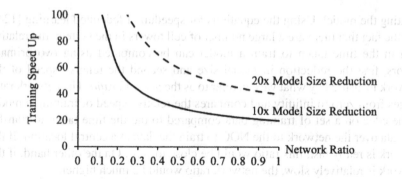
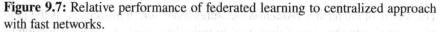

Figure 9.7: Relative performance of federated learning to centralized approach with fast networks.

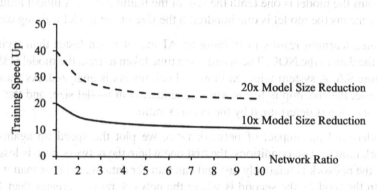
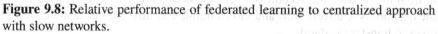

Figure 9.8: Relative performance of federated learning to centralized approach with slow networks.

shows a significant gain in the speed up from using federated learning. The reason is a significant increase in the parallelism of creating the model. Each of the cell towers trains its model on a segment of the data in parallel, which is responsible for the speed-up. In the case of network communications, where the cell towers range in tens of thousands, the speed up is tremendous.

Figure 9.8 shows the speed up when the network is relatively slow compared to the task of training a model. In this case, the speed up of federated learning is governed by the amount of data that needs to be transferred by the network. As a result, the speed up reaches the asymptotic limit of how much smaller the model is in comparison to the training data. Since models tend to be much smaller than the amount of data they are trained on, this results in federated learning-based approaches working much better and faster than the approach of learning models centrally.

9.3 Retail Coupon Recommendation

A case of federated learning arises in the retail industry which is motivated by security concerns instead of poor performance concerns like the use-case described in Section 9.2.

There have been several high profile thefts of consumer data from retailers [125, 126] and retail is the industry which has the largest incident of data breaches across multiple industry sectors [127]. Most retailers implement stringent data protection policies and strong IT security measures, but despite these, data breaches compromising user data happen due to human negligence or a break-down in the compliance procedures.

A major retailer in the United States was concerned that any potential data breach led to loss of significant amount of consumer data since all of the data was maintained in a central location. Instead of maintaining data in a central location, they were interested in letting the data remain at each retail store following the same security policies which were enforced in the central data warehouse. Customer purchase data at stores would be encrypted and stored locally without carrying the information to a central location. In case of any data breach, the compromised data would belong to only one store, which would be less expensive to handle and manage than a data breach at the central location which could expose information about customers across the entire chain. The underlying belief was that the data maintenance procedures at the retail chain were robust enough so that the probability of a data breach was not increased by such distribution.

The challenge in maintaining the data distributed in this manner is that the current data analysis and AI model learning infrastructure for the retail chain was built around the premise that all data was stored centrally. Therefore, equivalent mechanisms which would allow the same models to be built with distributed data needed to be developed. While there were several types of data analysis functions for a retail chain that can be implemented using distributed data, the pilot engagement was to demonstrate generation of retail coupons while maintaining distributed data as an example for other types of analysis.

Retail coupons are coupons that are given to shoppers when they check out their purchase at the cashier counter. These coupons provide an incentive for the shopper to buy targeted products on their next visit. The generation of these coupons is usually done by means of a top-N recommendation system [128] which determine which are the best items to recommend based on the transaction history of a customer. The retailer we were working with was using a variant of the SLIM algorithm [129] to generate its coupons.

The SLIM class of algorithms work by computing a matrix computing propensity to buy scores where the rows of the matrix are the groups of users and the columns of the products are the groups of products. The groups of users and products are

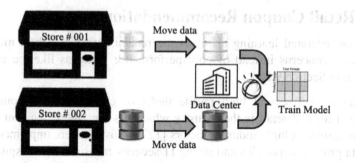

Figure 9.9: Centralized learning approach for generating recommendation models.

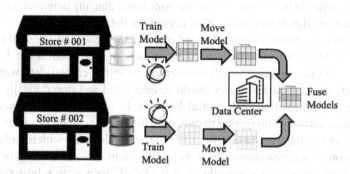

Figure 9.10: Federated learning approach for generating recommendation models.

computed by means of clustering of transactions so that users with the same buying behavior can be clustered into similar groups along rowss, and products with similar sales patterns are put into similar groups along columns. After the clustering is performed, the user group's propensity to buy from a group can be computed as a matrix. Coupons are generated by looking at the user current transaction, finding the group of products that the user is likely to buy from the computed matrix, and then selecting a product from the group not purchased by the user as an incentive for them to buy that product in the future.

The existing approach for generating coupons is to collect data to a central location, e.g. the retailer data center, and train the model on the collected data as shown in Figure 9.9. In order to generate the same model in a federated manner shown in Figure 9.10, one needs to generate the matrix for recommendation locally and then combine the recommendation matrices together. The combination of the matrices is a relatively simple composition as long as the groups or clusters for both the users and products are consistent across stores. However, the clustering done independently at different retail stores need not be consistent.

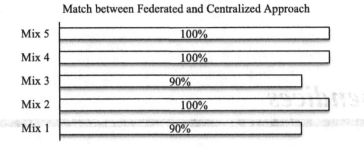

Figure 9.11: Results on generating recommendation models.

In order to make the clustering of the product groups and user groups consistent across all sites, we break the task of computing the recommendation matrix in two stages, the first being one in which users and product group clusters are identified. This effectively results in a federated clustering algorithm. Since clustering is a general function modeled by neural networks, it can be done using the algorithms described in Chapter 3 or it can be done with distributed algorithms specifically designed with clustering in mind [130, 131]. Once the clusters are consistent across all sites, each site can learn its matrix, and the aggregated matrix is a weighted average of those matrices. This process is both efficient and scales well with size of matrix, number of stores and and number of users.

A comparison of the recommendations offered by the two different approaches was done using data representative of retail transactions. Ten different buying patterns among different users were simulated at two different stores, and different mixes of the patterns at stores were run through both types of coupon recommendation system, one where the coupon recommendation matrix was learned in a centralized manner, and one in which it was learned in a federated manner. The results are shown in Figure 9.11. The overall match indicates whether products in the same group were recommended by both of the algorithms. The overall match of 94% indicates that the algorithms are effectively consistent with each other, and maintaining distributed data does not impact the ability to build AI models.

9.4 Summary

In this section, we have examined in brief some of the use-cases of federated learning applied to business use-cases. While these three are exemplar use-cases, federated learning and federated inference can be used in many other situations that are analogous to the ones described above.

The three use cases show the diversity of applications of federated learning. When used in combination with inference as well as learning, federated AI can provide viable solutions to several other challenges in modern business environments.

Appendices

Appendix 1: Frameworks for Federated Learning

Any federated learning system needs to create a software to implement the federated learning algorithms. The framework provides a basic infrastructure around which federated learning solutions can be built. The framework needs to be consistent with the overall approach for training AI/ML within the enterprise. The framework software would also support an interface which would allow the software developers to invoke different existing functions for federated learning functions, communicate across nodes involved in federated learning, and provide some libraries of existing algorithms.

An example framework for federated learning is federated tensorflow [132] (available at url https://www.tensorflow.org/federated). This framework is open-source and offered by Google researchers. It assumes that a neural network system based on Google tensorflow and the default mode of training is being done on Android phones.

Another framework geared for enterprise computing is offered by IBM [133] (available at urlhttps://ibmfl.mybluemix.net). It is available as a community edition which is open-sourced and is an extension to IBM commercial products dealing with distributed data. The community framework provides a library of federated learning algorithms which include federated learning across neural networks, decision trees and several other algorithms. As a library of python functions, it is geared more towards machines running federated learning routines as opposed to support on mobile phones.

Baidu, the leading search engine in China, has also published its framework for federated learning as an extension to its broader framework for AI called PaddlePaddle (available at https://github.com/PaddlePaddle). The larger PaddlePaddle is an alternative to learning frameworks like tensorflow or PyTorch, and since middle of 2000 include the tools that support federated averaging support. PaddlePaddle is geared towards learning solutions in industrial environments.

In addition to the above frameworks published by large companies, the Flower framework [134] (available at https://flower.dev) has been published by researchers from European Universities. The flower framework is targeted at researchers creating new algorithms and extending federated learning to many new different environments.

Given the significance and importance of federated learning, more such frameworks are likely to emerge. Organizations implementing federated learning-based solutions may opt to choose one of the existing frameworks, or implement their own framework for federated learning. The choice of the approach would depend on factors such as the closeness to existing approaches for AI/ML within the organization, any licensing considerations, and ease of use. As an example, organizations relying on products from IBM to run their functions might find the IBM framework more suitable for their needs, while companies in China may prefer PaddlePaddle. Other organizations may find it easier to use the Google framework or to develop their own federated learning approaches on a completely new AI framework, e.g. a new framework on PySoft.

Appendix 2: Adversarial Federated Learning

The main focus of the book has been in the context of businesses that have data distributed among many different sites and need to learn an AI model without necessarily moving data to a single location. In Section 2.3.1 of Chapter 2, we had briefly introduced the concept of consumer federated learning. Within the context of consumer federated learning, adversarial learning is an important concern.

Consumer federated learning referred to the scenario where several mobile phones were maintaining data about their user which was mined and analyzed locally. Consumer federated learning has been the focus of intense academic research and has resulted in several publications [30, 31, 135]. Example applications of such consumer-focused federated learning include predicting keyboard type-ahead or predicting the set of user browser queries. The motivation is usually protection of user privacy in the sense that private data belonging to the user need not move outside the mobile phone.

Consumer federated learning has the advantage that all learning is done for a specific mobile application which can have a fixed format of input over which the models are being trained, thereby avoiding the complexities of the algorithms described in Chapters 4 and 7. The learning can be coordinated to happen in a synchronized manner, avoiding the issues and approaches discussed in Chapter 7. Thus, the systems implementing consumer federated learning can focus on creating the model and using it in the context of the specific mobile application.

The challenges associated with consumer federated learning have to deal with a significant large number of end points with data, which results in each mobile phone having data that is very specific to the user. Unfortunately, this also means that the data at each node is too small to make a meaningful model in most cases. In order to work around this lack of data, cohorts of users are used to combine data from several users into a larger group where sufficient data can be used to create a useful model [136], but this means that data has to be moved out of the mobile phone to a set of cohorts in order to create the model. This either requires complex security protocols to maintain privacy, like secure multi-party computation [137], that are very inefficient, or else violate the privacy requirement that was the original motivation for federated learning. Another challenge is the existence of malicious parties which can launch adverserial attacks on federated learning.

The current prevalent practice in the industry is to work around the privacy requirements of consumer data on the mobile phone using business arrangements which require users to agree to the use of their data in exchange for the benefit delivered by the mobile application. As a result, the primary motivation for consumer federated learning can be eliminated with simple business mechanisms, and the approaches are mostly of academic interest. While the high level of academic interest is shown in the large number of papers published on this subject, federated learning on mobile phones for consumers is likely to stay within the realm of academic research and unlikely to see wide adoption in the enterprise. Businesses are more likely to adopt a business mechanism to collect the data centrally and build AI models from them, which provides a more simplified and traditional implementation of AI-based systems.

Adversarial federated learning is an important challenge associated with consumer federated learning. When several consumer devices are generating data and that data is being used to train a common model, there is a high probability that someone may be feeding bad data or bad models into the system. The number of ways in which the common model being trained together can be attacked in a variety of ways, not just by one single participant but also by a set of maliciously coordinated cohort of clients.

A variety of attacks that can be targeted at different forms of federated learning can been described [138, 139, 140], which can create challenges for creating good models. In a massively distributed deployment of federate learning, it would be reasonable to expect that there would be pranksters and malicious entities, which is another reason for suspecting that widespread consumer federated learning is not likely to be used within enterprise applications anytime shortly. Such malicious entities can carefully select the weights they provide at different stages of the federated model training, leading to corruption of the model that is eventually created.

There have been approaches proposed which would mitigate the impact of such poisoning attacks, which include adjusting the averaging process of weights into it done in a hierarchical manner with a randomly selected groups of clients [141, 141]. This will reduce the impact of a malicious individual and randomize the malicious cohort members into different groups which should prevent the attacks if the malicious entities are a small fraction of the overall participants. Another safeguard can be used by keeping the individual parameters of the model being trained hidden from the other participants, who can only see the aggregated model. Since adverserial attacks need to be crafted carefully based on the type of models from other clients providing their models, putting in the privacy mechanisms can prevent targeted attacks against federated learning systems [142].

Other traditional mechanisms for detecting and preventing malicious parties, e.g. reputation-based schemes [67] and using mechanisms like differential privacy [104], secure multiparty computations [137] also provide some resilience against malicious entities.

We have not focused on this aspect within this book since this is a vulnerability which is typical for consumer federated learning but not likely to arise in the context of enterprise federated learning. In the context of enterprise federated learning, business arrangements provide a sufficient disincentive for anyone to participate willingly in a malicious adversarial attack on the federated learning process. While some inadvertent adverserial attacks may happen if software of one of the participants is compromised, the mechanisms for handling the variety of trust situations in Chapter 6 can provide sufficient safeguards for those situations.

References

[1] S. Legg, M. Hutter, *et al.*, "A collection of definitions of intelligence," in *Frontiers in Artificial Intelligence and Applications*, vol. 157, p. 17, IOS Press, 2007.

[2] C. Grosan and A. Abraham, "Rule-based expert systems," in *Intelligent Systems*, pp. 149–185, 2011.

[3] G. Cirincione and D. Verma, "Federated Machine Learning for Multi-Domain Operations at the tactical edge," in *Artificial Intelligence and Machine Learning for Multi-Domain Operations Applications*, vol. 11006, p. 1100606, International Society for Optics and Photonics, 2019.

[4] W. D. Nothwang, M. J. McCourt, R. M. Robinson, S. A. Burden, and J. W. Curtis, "The human should be part of the control loop?," in *2016 Resilience Week (RWS)*, pp. 214–220, IEEE, 2016.

[5] G. A. Seber and A. J. Lee, *Linear Regression Analysis*, vol. 329. John Wiley & Sons, 2012.

[6] D. G. Kleinbaum, K. Dietz, M. Gail, M. Klein, and M. Klein, *Logistic Regression*. Springer, 2002.

[7] L. Atzori, A. Iera, and G. Morabito, "The internet of things: A survey," *Computer Networks*, vol. 54, no. 15, pp. 2787–2805, 2010.

[8] R. Kohavi, "The power of decision tables," in *European Conference on Machine Learning*, pp. 174–189, Springer, 1995.

[9] L. Rokach and O. Maimon, "Decision trees," in *Data Mining and Knowledge Discovery Handbook*, pp. 165–192, Springer, 2005.

[10] C. Feng and D. Michie, "Machine learning of rules and trees," *Machine Learning, Neural and Statistical Classification*, pp. 50–83, 1994.

[11] M. A. Nielsen, *Neural Networks and Deep Learning*, vol. 2018. Determination Press San Francisco, CA, 2015.

[12] C. C. Aggarwal, *Neural Networks and Deep Learning*. Springer, 2018.

[13] H. Abdi and L. J. Williams, "Principal component analysis," *Wiley Interdisciplinary Reviews: Computational Statistics*, vol. 2, no. 4, pp. 433–459, 2010.

[14] H. Abdi and D. Valentin, "Multiple correspondence analysis," *Encyclopedia of Measurement and Statistics*, vol. 2, no. 4, pp. 651–657, 2007.

[15] S. Suthaharan, "Support vector machine," in *Machine Learning Models and Algorithms for Big Data Classification*, pp. 207–235, Springer, 2016.

[16] B. Steffen, F. Howar, and M. Merten, "Introduction to active automata learning from a practical perspective," in *International School on Formal Methods for the Design of Computer, Communication and Software Systems*, pp. 256–296, Springer, 2011.

[17] J. G. Kemeny and J. L. Snell, *Markov Chains*. Springer-Verlag, New York, 1976.

[18] L. Rabiner and B. Juang, "An introduction to Hidden Markov Models," *IEEE ASSP Magazine*, vol. 3, no. 1, pp. 4–16, 1986.

[19] T. Murata, "Petri Nets: Properties, Analysis and Applications," *Proceedings of the IEEE*, vol. 77, no. 4, pp. 541–580, 1989.

[20] S. E. Schaeffer, "Graph clustering," *Computer Science Review*, vol. 1, no. 1, pp. 27–64, 2007.

[21] T. Velte, A. Velte, and R. Elsenpeter, *Cloud Computing, A Practical Approach*. McGraw-Hill, Inc., 2009.

[22] M. J. Wolf, K. W. Miller, and F. S. Grodzinsky, "Why we should have seen that coming: comments on microsoft's tay "experiment," and wider implications," *The ORBIT Journal*, vol. 1, no. 2, pp. 1–12, 2017.

[23] C. Szegedy, W. Zaremba, I. Sutskever, J. Bruna, D. Erhan, I. Goodfellow, and R. Fergus, "Intriguing properties of neural networks," *arXiv preprint arXiv:1312.6199*, 2013.

[24] N. Carlini and D. Wagner, "Audio adversarial examples: Targeted attacks on speech-to-text," in *2018 IEEE Security and Privacy Workshops (SPW)*, pp. 1–7, IEEE, 2018.

[25] W. Y. B. Lim, N. C. Luong, D. T. Hoang, Y. Jiao, Y.-C. Liang, Q. Yang, D. Niyato, and C. Miao, "Federated learning in mobile edge networks: A comprehensive survey," *IEEE Communications Surveys & Tutorials*, 2020.

[26] T. Nelson, *Mergers and Acquisitions from A to Z*. Amacom, 2018.

[27] V. G. da Silva, M. Kirikova, and G. Alksnis, "Containers for virtualization: An overview," *Applied Computer Systems*, vol. 23, no. 1, pp. 21–27, 2018.

[28] M. Ramsay, O. Sankoh, as members of the AWI-Gen study, and the H3Africa Consortium, "African partnerships through the H3Africa Consortium bring a genomic dimension to longitudinal population studies on the continent," 2016.

[29] A. C. Ezeh, C. O. Izugbara, C. W. Kabiru, S. Fonn, K. Kahn, L. Manderson, A. S. Undieh, A. Omigbodun, and M. Thorogood, "Building capacity for public and population health research in Africa: the consortium for advanced research training in Africa (CARTA) model," *Global Health Action*, vol. 3, no. 1, p. 5693, 2010.

[30] H. B. McMahan, E. Moore, D. Ramage, S. Hampson, *et al.*, "Communication-efficient learning of deep networks from decentralized data," *International Conference on Artificial Intelligence and Statistics (AISTATS)*, 2017.

[31] K. Bonawitz, V. Ivanov, B. Kreuter, A. Marcedone, H. B. McMahan, S. Patel, D. Ramage, A. Segal, and K. Seth, "Practical secure aggregation for federated learning on user-held data," *arXiv preprint arXiv:1611.04482*, 2016.

[32] E. A. Lile, "Client/Server architecture: A brief overview," *Journal of Systems Management*, vol. 44, no. 12, p. 26, 1993.

[33] J. Dean and S. Ghemawat, "MapReduce: simplified data processing on large clusters," *Communications of the ACM*, vol. 51, no. 1, pp. 107–113, 2008.

[34] T. Hofmann, B. Schölkopf, and A. J. Smola, "Kernel methods in machine learning," *The Annals of Statistics*, pp. 1171–1220, 2008.

[35] M. V. Wickerhauser, *Adapted Wavelet Analysis: From Theory to Software.* CRC Press, 1996.

[36] D. A. Reynolds, "Gaussian mixture models," *Encyclopedia of Biometrics*, vol. 741, 2009.

[37] J. Konečný, H. B. McMahan, F. X. Yu, P. Richtárik, A. T. Suresh, and D. Bacon, "Federated learning: Strategies for improving communication efficiency," *arXiv preprint arXiv:1610.05492*, 2016.

[38] S. Wang, T. Tuor, T. Salonidis, K. K. Leung, C. Makaya, T. He, and K. Chan, "When edge meets learning: Adaptive control for resource-constrained distributed machine learning," in *IEEE INFOCOM 2018-IEEE Conference on Computer Communications*, pp. 63–71, IEEE, 2018.

[39] E. Bakopoulou, B. Tillman, and A. Markopoulou, "A federated learning approach for mobile packet classification," *arXiv preprint arXiv:1907.13113*, 2019.

[40] Y. Liang, M.-F. F. Balcan, V. Kanchanapally, and D. Woodruff, "Improved distributed principal component analysis," in *Advances in Neural Information Processing Systems*, pp. 3113–3121, 2014.

[41] L. Rokach and O. Maimon, "Top-down induction of decision trees classifiers-a survey," *IEEE Transactions on Systems, Man, and Cybernetics, Part C (Applications and Reviews)*, vol. 35, no. 4, pp. 476–487, 2005.

[42] J. D. Moffett and M. S. Sloman, "Policy conflict analysis in distributed system management," *Journal of Organizational Computing and Electronic Commerce*, vol. 4, no. 1, pp. 1–22, 1994.

[43] A. Nedic and D. P. Bertsekas, "Incremental subgradient methods for non-differentiable optimization," *SIAM Journal on Optimization*, vol. 12, no. 1, pp. 109–138, 2001.

[44] N. Shan and W. Ziarko, "An incremental learning algorithm for constructing decision rules," in *Rough Sets, Fuzzy Sets and Knowledge Discovery*, pp. 326–334, Springer, 1994.

[45] L. Guan, "An incremental updating algorithm of attribute reduction set in decision tables," in *Sixth International Conference on Fuzzy Systems and Knowledge Discovery*, vol. 2, pp. 421–425, IEEE, 2009.

[46] P. E. Utgoff, "Incremental Induction of Decision Trees," *Machine Learning*, vol. 4, no. 2, pp. 161–186, 1989.

[47] S. Ruping, "Incremental learning with support vector machines," in *Proceedings 2001 IEEE International Conference on Data Mining*, pp. 641–642, IEEE, 2001.

[48] D. Opitz and R. Maclin, "Popular ensemble methods: An empirical study," *Journal of Artificial Intelligence Research*, vol. 11, pp. 169–198, 1999.

[49] T. Bray (Ed.), "The JavaScript Object Notation (JSON) Data Interchange Format." RFC 8259 (Internet Standard), Dec. 2017.

[50] T. Bray, J. Paoli, C. M. Sperberg-McQueen, E. Maler, F. Yergeau, *et al.*, "Extensible markup language (XML) 1.0," 2000.

[51] M. Nixon and A. Aguado, *Feature Extraction and Image Processing for Computer Vision*. Academic Press, 2019.

[52] L. Rabiner and R. Schafer, *Theory and Applications of Digital Speech Processing*. Prentice Hall Press, 2010.

[53] C. Brodie, D. George, C.-M. Karat, J. Karat, J. Lobo, M. Beigi, X. Wang, S. Calo, D. Verma, A. Schaeffer-Filho, *et al.*, "The coalition policy management portal for policy authoring, verification, and deployment," in *2008 IEEE Workshop on Policies for Distributed Systems and Networks*, pp. 247–249, IEEE, 2008.

[54] S. KIKUCHI, Y. KANNA, and Y. ISOZAKI, "CIM Simplified Policy Language (CIM-SPL) CIM Simplified Policy Language (CIM-SPL), 2009," *IEICE Transactions on Information and Systems*, vol. 95, no. 11, pp. 2634–2650, 2012.

[55] S.-L. Documentation, "Compare the effect of different scalers on data with outliers," Nov 2020.

[56] L. He, L.-d. WU, and Y.-c. CAI, "Survey of clustering algorithms in data mining," *Application Research of Computers*, vol. 1, pp. 10–13, 2007.

[57] C. Bisdikian, L. M. Kaplan, M. B. Srivastava, D. J. Thornley, D. Verma, and R. I. Young, "Building principles for a quality of information specification for sensor information," in *2009 12th International Conference on Information Fusion*, pp. 1370–1377, IEEE, 2009.

[58] A. Bar-Noy, G. Cirincione, R. Govindan, S. Krishnamurthy, T. LaPorta, P. Mohapatra, M. Neely, and A. Yener, "Quality of information aware networking for tactical military networks," in *IEEE International Conference on Pervasive Computing and Communications*, 2011.

[59] V. Sachidananda, A. Khelil, and N. Suri, "Quality of information in wireless sensor networks: A survey," in *Proceeding of the International Conference on Information Quality*, 2010.

[60] H. Jin, L. Su, D. C. K. Nahrstedt, and J. Xu, "Quality of information aware incentive mechanisms for mobile crowd sensing systems," in *Proceedings of the ACM International Symposium on Mobile Ad-Hoc Networking and Computing*, pp. 167–176, 2015.

[61] D. Turgut and L. Boloni, "Value of information and cost of privacy in the internet of things," *IEEE Communications Magazine*, vol. 55, no. 9, pp. 62–66, 2017.

[62] F. A. Khan, S. Butt, S. A. Khan, L. Bölöni, and D. Turgut, "Value of information based data retrieval in uwsns," *Sensors*, vol. 18, no. 10, p. 3414, 2018.

[63] D. Verma, G. de Mel, and G. Pearson, "VoI for complex AI based solutions in coalition environments," in *Artificial Intelligence and Machine Learning for Multi-Domain Operations Applications*, vol. 11006, p. 1100609, International Society for Optics and Photonics, 2019.

[64] K. Grueneberg, S. Calo, P. Dewan, D. Verma, and T. O'Gorman, "A Policy-based Approach for Measuring Data Quality," in *2019 IEEE International Conference on Big Data (Big Data)*, pp. 4025–4031, IEEE, 2019.

[65] D. Verma, S. Calo, S. Witherspoon, I. Manotas, E. Bertino, A. A. Jabal, G. de Mel, A. Swami, G. Cirincione, and G. Pearson, "Managing training data from untrusted partners using self-generating policies," in *Artificial Intelligence and Machine Learning for Multi-Domain Operations Applications*, vol. 11006, p. 110060P, International Society for Optics and Photonics, 2019.

[66] D. Verma, S. Calo, S. Witherspoon, I. Manotas, E. Bertino, A. M. A. Jabal, G. Cirincione, A. Swami, G. Pearson, and G. de Mel, "Self-generating policies for machine learning in coalition environments," in *Policy-Based Autonomic Data Governance*, pp. 42–65, Springer, 2019.

[67] F. Hendrikx, K. Bubendorfer, and R. Chard, "Reputation systems: A survey and taxonomy," *Journal of Parallel and Distributed Computing*, vol. 75, pp. 184–197, 2015.

[68] G. Pearson, D. Verma, and G. de Mel, "Value of information: Quantification and application to coalition machine learning," in *Policy-Based Autonomic Data Governance*, pp. 21–41, Springer, 2019.

[69] D. Verma, S. Calo, S. Witherspoon, E. Bertino, A. A. Jabal, A. Swami, G. Cirincione, S. Julier, G. White, G. de Mel, *et al.*, "Federated Learning for Coalition Operations," in *AAAI Fall Symposium Series: Artificial Intelligence in Government and Public Sector, Arlington, Virginia, USA*, AAAI, 2019.

[70] D. Verma, S. Julier, and G. Cirincione, "Federated AI for building ai solutions across multiple agencies," in *AAAI Fall Symposium Series: Artificial Intelligence in Government and Public Sector, Arlington, Virginia, USA*, AAAI, 2018.

[71] D. C. Verma, G. White, S. Julier, S. Pasteris, S. Chakraborty, and G. Cirincione, "Approaches to address the data skew problem in federated learning," in *Artificial Intelligence and Machine Learning for Multi-Domain Operations Applications*, vol. 11006, p. 110061I, International Society for Optics and Photonics, 2019.

[72] D. C. Verma, "Simplifying network administration using policy-based management," *IEEE Network*, vol. 16, no. 2, pp. 20–26, 2002.

[73] W. Han and C. Lei, "A survey on policy languages in network and security management," *Computer Networks*, vol. 56, no. 1, pp. 477–489, 2012.

[74] D. Verma, S. Calo, E. Bertino, A. Russo, and G. White, "Policy based Ensembles for applying ML on big data," in *2019 IEEE International Conference on Big Data (Big Data)*, pp. 4038–4044, IEEE, 2019.

[75] D. C. Verma, E. Bertino, A. Russo, S. Calo, and A. Singla, "Policy based ensembles for multi domain operations," in *Artificial Intelligence and Machine Learning for Multi-Domain Operations Applications II*, vol. 11413, p. 114130A, International Society for Optics and Photonics, 2020.

[76] M. Fredrikson, S. Jha, and T. Ristenpart, "Model inversion attacks that exploit confidence information and basic countermeasures," in *Proceedings of the 22nd ACM SIGSAC Conference on Computer and Communications Security*, pp. 1322–1333, 2015.

[77] A. Creswell, T. White, V. Dumoulin, K. Arulkumaran, B. Sengupta, and A. A. Bharath, "Generative adversarial networks: An overview," *IEEE Signal Processing Magazine*, vol. 35, no. 1, pp. 53–65, 2018.

[78] B. Hitaj, G. Ateniese, and F. Perez-Cruz, "Deep models under the GAN: information leakage from collaborative deep learning," in *Proceedings of the 2017 ACM SIGSAC Conference on Computer and Communications Security*, pp. 603–618, 2017.

[79] G. Pearson and T. Pham, "The challenge of sensor information processing and delivery within network and information science research," in *Defense Transformation and Net-Centric Systems*, vol. 6981, p. 698105, International Society for Optics and Photonics, 2008.

[80] T. Pham, G. Cirincione, A. Swami, G. Pearson, and C. Williams, "Distributed Analytics and Information Sciences," in *International Conference on Information Fusion*, pp. 245–252, IEEE, 2015.

[81] K. E. Schaefer, J. Oh, D. Aksaray, and D. Barber, "Integrating Context into Artificial Intelligence: Research from the Robotics Collaborative Technology Alliance," *AI Magazine*, vol. 40, no. 3, pp. 28–40, 2019.

[82] S. Wolford, *The Politics of Military Coalitions*. Cambridge University Press, 2015.

[83] T. Pham, G. H. Cirincione, D. Verma, and G. Pearson, "Intelligence, Surveillance and Reconnaissance fusion for Coalition Operations," in *International Conference on Information Fusion*, pp. 1–8, IEEE, 2008.

[84] P. R. Smart and K. P. Sycara, "Collective Sensemaking and Military Coalitions," *IEEE Intelligent Systems*, vol. 28, no. 1, pp. 50–56, 2012.

[85] R. Rathmell, "A Coalition Force Scenario'Binni-Gateway to the Golden Bowl of Africa'," in *Proceedings of the International Workshop on Knowledge-based Planning for Coalition Forces (ed. Tate, A.)*, pp. 115–125, 1999.

[86] D. Roberts, G. Lock, and D. C. Verma, "Holistan: A futuristic scenario for international coalition operations," in *2007 International Conference on Integration of Knowledge Intensive Multi-Agent Systems*, pp. 423–427, IEEE, 2007.

[87] R. R. Henning, "Security service level agreements: quantifiable security for the enterprise?," in *Proceedings of the 1999 Workshop on New Security Paradigms*, pp. 54–60, 1999.

[88] L. Wu and R. Buyya, "Service Level Agreement (SLA) in utility computing systems," in *Performance and Dependability in Sservice Computing: Concepts, Techniques and Research directions*, pp. 1–25, IGI Global, 2012.

[89] L. Karadsheh, "Applying security policies and service level agreement to IaaS service model to enhance security and transition," *Computers & Security*, vol. 31, no. 3, pp. 315–326, 2012.

[90] D. C. Verma, "Service level agreements on IP networks," *Proceedings of the IEEE*, vol. 92, no. 9, pp. 1382–1388, 2004.

[91] K. Sarpatwar, R. Vaculin, H. Min, G. Su, T. Heath, G. Ganapavarapu, and D. Dillenberger, "Towards enabling trusted artificial intelligence via blockchain," in *Policy-Based Autonomic Data Governance*, pp. 137–153, Springer, 2019.

[92] I. Bashir, *Mastering Blockchain: Distributed ledger technology, decentralization, and smart contracts explained.* Packt Publishing Ltd, 2018.

[93] R. Yegireddi and R. K. Kumar, "A survey on conventional encryption algorithms of Cryptography," in *2016 International Conference on ICT in Business Industry & Government (ICTBIG)*, pp. 1–4, IEEE, 2016.

[94] B. Kaliski, "A survey of encryption standards," *IEEE Micro*, vol. 13, no. 6, pp. 74–81, 1993.

[95] C. Gentry, "Fully homomorphic encryption using ideal lattices," in *Proceedings of the Forty-first Annual ACM Symposium on Theory of Computing*, pp. 169–178, 2009.

[96] Z. Brakerski, C. Gentry, and V. Vaikuntanathan, "Fully homomorphic encryption without bootstrapping," *ACM Transactions on Computation Theory (TOCT)*, vol. 6, no. 3, pp. 1–36, 2014.

[97] M.-q. Hong, P.-Y. Wang, and W.-B. Zhao, "Homomorphic encryption scheme based on elliptic curve cryptography for privacy protection of cloud computing," in *2016 IEEE 2nd International Conference on Big Data Security on Cloud (BigDataSecurity), IEEE International Conference on High Performance and Smart Computing (HPSC), and IEEE International Conference on Intelligent Data and Security (IDS)*, pp. 152–157, IEEE, 2016.

[98] S. S. Roy, F. Turan, K. Jarvinen, F. Vercauteren, and I. Verbauwhede, "FPGA-based high-performance parallel architecture for homomorphic computing on encrypted data," in *2019 IEEE International Symposium on High Performance Computer Architecture (HPCA)*, pp. 387–398, IEEE, 2019.

[99] A. Al Badawi, B. Veeravalli, J. Lin, N. Xiao, M. Kazuaki, and A. K. M. Mi, "Multi-GPU design and performance evaluation of homomorphic encryption on GPU clusters," *IEEE Transactions on Parallel and Distributed Systems*, vol. 32, no. 2, pp. 379–391, 2020.

[100] E. Milanov, "The RSA algorithm," *RSA Laboratories*, pp. 1–11, 2009.

[101] T. ElGamal, "A public key cryptosystem and a signature scheme based on discrete logarithms," *IEEE Transactions on Information Theory*, vol. 31, no. 4, pp. 469–472, 1985.

[102] P. Paillier, "Public-key cryptosystems based on composite degree residuosity classes," in *International Conference on the Theory and Applications of Cryptographic Techniques*, pp. 223–238, Springer, 1999.

[103] S. Chakraborty, C. Liu, and D. Verma, "Secure model fusion for distributed Learning using partial homomorphic encryption," in *Policies for Autonomic Data Governance at ESORICS*, 2018.

[104] C. Dwork, "Differential privacy: A survey of results," in *International Conference on Theory and Applications of Models of Computation*, pp. 1–19, Springer, 2008.

[105] V. Pimentel and B. G. Nickerson, "Communicating and displaying real-time data with websocket," *IEEE Internet Computing*, vol. 16, no. 4, pp. 45–53, 2012.

[106] M. McCloskey and N. J. Cohen, "Catastrophic interference in connectionist networks: The sequential learning problem," in *Psychology of Learning and Motivation*, vol. 24, pp. 109–165, Elsevier, 1989.

[107] W. J. Scheirer, A. de Rezende Rocha, A. Sapkota, and T. E. Boult, "Toward open set recognition," *IEEE Transactions on Pattern Analysis and Machine Intelligence*, vol. 35, no. 7, pp. 1757–1772, 2012.

[108] W. J. Scheirer, L. P. Jain, and T. E. Boult, "Probability models for open set recognition," *IEEE Transactions on Pattern Analysis and Machine Intelligence*, vol. 36, no. 11, pp. 2317–2324, 2014.

[109] F. Le, K. Leung, K. Poularakis, L. Tassiulas, and Y. Paul, "Extracting Interpretable Rules from Deep Models across Coalitions," in *Proceedings of the Annual Fall Meeting of DAIS ITA*, 2020.

[110] D. Bahdanau, K. Cho, and Y. Bengio, "Neural machine translation by jointly learning to align and translate," in *3rd International Conference on Learning Representations, ICLR 2015*, 2015.

[111] V. Mnih, N. Heess, A. Graves, *et al.*, "Recurrent models of visual attention," in *Advances in Neural Information Processing Systems*, pp. 2204–2212, 2014.

[112] J. Quinlan, "Generating production rules from decision trees," in *Proceedings of the 10th International Joint Conference on Artificial Intelligence*, pp. 304–307, Morgan Kaufmann Publishers Inc., 1987.

[113] L. O. Hall, N. Chawla, K. W. Bowyer, *et al.*, "Combining decision trees learned in parallel," in *Knowledge Discovery and Data Mining Workshop on Distributed Data Mining*, pp. 10–15, 1998.

[114] D. C. Verma, *Policy-based Networking: Architecture and Algorithms.* New Riders Publishing, 2000.

[115] M.-F. F. Balcan, S. Ehrlich, and Y. Liang, "Distributed k-means and k-median clustering on general topologies," *Advances in Neural Information Processing Systems*, vol. 26, pp. 1995–2003, 2013.

[116] O. Bachem, M. Lucic, and A. Krause, "Practical coreset constructions for machine learning," *arXiv preprint arXiv:1703.06476*, 2017.

[117] H. Lu, M.-J. Li, T. He, S. Wang, V. Narayanan, and K. S. Chan, "Robust coreset construction for distributed machine learning," *IEEE Journal on Selected Areas in Communications*, vol. 38, no. 10, pp. 2400–2417, 2020.

[118] D. W. Cheung, V. T. Ng, A. W. Fu, and Y. Fu, "Efficient mining of association rules in distributed databases," *IEEE Transactions on Knowledge and Data Engineering*, vol. 8, no. 6, pp. 911–922, 1996.

[119] B. Gu, Z. Dang, X. Li, and H. Huang, "Federated doubly stochastic kernel learning for vertically partitioned data," in *Proceedings of the 26th ACM SIGKDD International Conference on Knowledge Discovery & Data Mining*, pp. 2483–2493, 2020.

[120] Y. Kang, Y. Liu, and T. Chen, "FedMVT: Semi-supervised vertical federated learning with multiView training," *IJCAI Workshop on Federated Learning for User Privacy and Data Confidentiality*, 2020.

[121] P. Voigt and A. Von dem Bussche, "The eu general data protection regulation (gdpr)," *A Practical Guide, 1st Ed., Cham: Springer International Publishing*, 2017.

[122] P. Richhariya and P. K. Singh, "A survey on financial fraud detection methodologies," *International Journal of Computer Applications*, vol. 45, no. 22, pp. 15–22, 2012.

[123] D. C. Verma and P. Verma, *Techniques for Surviving the Mobile Data Explosion.* John Wiley & Sons, 2014.

[124] D. Verma, S. Julier, and G. Cirincione, "Federated AI for building AI solutions across multiple agencies," in *AAAI FSS-18: Artificial Intelligence in Government and Public Sector, Arlington, VA, USA*, 2018.

[125] M. Plachkinova and C. Maurer, "Security breach at target," *Journal of Information Systems Education*, vol. 29, no. 1, pp. 11–20, 2018.

[126] E. Marseille, *The Rapid Growth of Data Breaches in Today's Society.* PhD thesis, Utica College, 2020.

[127] K. D. Harris and A. General, "California data breach report," *Retrieved August*, vol. 7, p. 2016, 2016.

[128] D. Paraschakis, B. J. Nilsson, and J. Hollander, "Comparative evaluation of top-n recommenders in e-commerce: An industrial perspective," in *IEEE 14th International Conference on Machine Learning and Applications (ICMLA)*, pp. 1024–1031, IEEE, 2015.

[129] X. Ning and G. Karypis, "SLIM: Sparse linear methods for top-n recommender systems," in *2011 IEEE 11th International Conference on Data Mining*, pp. 497–506, IEEE, 2011.

[130] H. Kargupta, W. Huang, K. Sivakumar, and E. Johnson, "Distributed clustering using collective principal component analysis," *Knowledge and Information Systems*, vol. 3, no. 4, pp. 422–448, 2001.

[131] E. Januzaj, H.-P. Kriegel, and M. Pfeifle, "Scalable density-based distributed clustering," in *European Conference on Principles of Data Mining and Knowledge Discovery*, pp. 231–244, Springer, 2004.

[132] K. Bonawitz, H. Eichner, W. Grieskamp, D. Huba, A. Ingerman, V. Ivanov, C. Kiddon, J. Konečný, S. Mazzocchi, H. B. McMahan, *et al.*, "Towards Federated Learning at Scale: System Design," *arXiv preprint arXiv:1902.01046*, 2019.

[133] H. Ludwig, N. Baracaldo, G. Thomas, Y. Zhou, A. Anwar, S. Rajamoni, Y. Ong, J. Radhakrishnan, A. Verma, M. Sinn, *et al.*, "IBM federated learning: an enterprise framework white paper v0. 1," *arXiv preprint arXiv:2007.10987*, 2020.

[134] D. J. Beutel, T. Topal, A. Mathur, X. Qiu, T. Parcollet, and N. D. Lane, "Flower: A Friendly Federated Learning Research Framework," *arXiv e-prints*, pp. arXiv–2007, 2020.

[135] M. Aledhari, R. Razzak, R. M. Parizi, and F. Saeed, "Federated learning: A survey on enabling technologies, protocols, and applications," *IEEE Access*, vol. 8, pp. 140699–140725, 2020.

[136] F. Sattler, K.-R. Müller, and W. Samek, "Clustered federated learning: Model-agnostic distributed multitask optimization under privacy constraints," *IEEE Transactions on Neural Networks and Learning Systems*, 2020.

[137] R. Cramer, I. B. Damgard, , and J. B. Nielsen, *Secure Multiparty Computation*. Cambridge University Press, 2015.

[138] A. N. Bhagoji, S. Chakraborty, P. Mittal, and S. Calo, "Analyzing federated learning through an adversarial lens," in *International Conference on Machine Learning*, pp. 634–643, PMLR, 2019.

[139] E. Bagdasaryan, A. Veit, Y. Hua, D. Estrin, and V. Shmatikov, "How to backdoor federated learning," in *International Conference on Artificial Intelligence and Statistics*, pp. 2938–2948, PMLR, 2020.

[140] J. Zhang, J. Chen, D. Wu, B. Chen, and S. Yu, "Poisoning attack in federated learning using generative adversarial nets," in *2019 18th IEEE International Conference on Trust, Security and Privacy in Computing and Communications/13th IEEE International Conference on Big Data Science and Engineering (TrustCom/BigDataSE)*, pp. 374–380, IEEE, 2019.

[141] Z. Chen, P. Tian, W. Liao, and W. Yu, "Zero Knowledge Clustering Based Adversarial Mitigation in Heterogeneous Federated Learning," *IEEE Transactions on Network Science and Engineering*, 2020.

[142] Y. Song, T. Liu, T. Wei, X. Wang, Z. Tao, and M. Chen, "FDA3: federated defense against adversarial attacks for cloud-based iiot applications," *IEEE Transactions on Industrial Informatics*, 2020.

Index